Gaponov-Grekhov · Rabinovich Nonlinearities in Action

A. V. Gaponov-Grekhov
M. I. Rabinovich

Nonlinearities in Action

Oscillations Chaos Order Fractals

With 121 Figures and 63 Color Plates

Springer-Verlag
Berlin Heidelberg New York
London Paris Tokyo
Hong Kong Barcelona
Budapest

Professor Andrei V. Gaponov-Grekhov
Member of the Russian Academy of Sciences

Professor Mikhail I. Rabinovich
Corresponding Member of the Russian Academy of Sciences

Institute of Applied Physics, ul. Ulyanova 46
603 600 Nizhny Novgorod, Russia

English by:

Dr. Ernst F. Hefter

Klosterstrasse 85
D-74931 Lobenfeld, Germany

Nadya Krivatkina

Institute of Applied Physics
Russian Academy of Sciences
ul. Ulyanova 46
603 600 Nizhny Novgorod, Russia

ISBN 3-540-51988-2 Springer-Verlag Berlin Heidelberg New York
ISBN 0-387-51988-2 Springer-Verlag New York Berlin Heidelberg

Library of Congress Cataloging-in-Publication Data. Gaponov-Grekhov, A. V. Nonlinearities in action : oscillations, chaos, order, fractals / A. V. Gaponov-Grekhov, M. I. Rabinovich. p. cm. Includes bibliographical references and index. ISBN 3-540-51988-2 (Berlin). – ISBN 0-387-51988-2 (New York) 1. Nonlinear theories. 2. Mathematical physics. I. Rabinovich, M. I. II. Title. QC20.7.N6G36 1992 530.1'5–dc20 92-8461

Printed in Hong Kong

Cover Design: Struve & Partner, Atelier für Grafik-Design, Heidelberg
Production Editor: I. Kaiser
Typesetting: Springer $\mathrm{T_EX}$ in-house system
57/3140 - 5 4 3 2 1 - Printed on acid-free paper

Foreword

In the dynamics of mankind one can trace out a path of contemplation about the "world", leading from early speculations to today's natural sciences. The endeavour to understand how nature works has led to the construction, still in progress, of an abstract building of great complexity. To the uninitiated it may look more like a scurrilous sculpture resting on many legs, among them such peculiar ones as probability, relativity, quantum mechanics... . At times problems with the stability of the building or sculpture arise: known facts that won't fit and can no longer be ignored start to undermine the foundations. Then new footings are thought of, constructed and finally cast. In fact, the undermining and casting is often done in one step.

This process has already been repeated many times and will undoubtedly repeat itself again and again. At present, one recognizable footing under construction goes by the name of "chaos theory". Physicists seem to like the word chaos. When they came to recognize that the air is not just empty space but an obviously indescribable dance of myriads of molecules they called that "chaos". What else would fit? In the course of time the name was simplified to "gas". Thus the word chaos became free to serve for the next upsetting experience. That arose in the context of nonlinear dynamical systems, where peculiar motions were detected, ones seemingly beyond human comprehension. Things like fractal dimensions had to be invented and much more! The investigations now leave no doubt that the peculiar findings stem from the fact that determinism does not imply predictability. Poincaré encountered this fact in connection with the three-body problem. But at that time, relativity and later quantum theory kept the scientists busy and prevented them from thinking about such a subtle question.

The old chaos theory gave way to kinetic gas theory and statistical mechanics. Where will the new chaos theory lead? Where will we find the tools to cope with the absence of predictability in most deterministic systems? Which deterministic systems are not affected? We are far from knowing the answers, but at the heart of this new chaos lies nonlinearity, the topic explored in depth in the present book. It can be said for sure that a huge edifice will be built on that footing under such names as "nonlinear physics", "nonlinear dynamics" and "nonlinear science".

It is always amusing to see scientists rushing into new territory with their different tools, each trying to dig and wash out treasures. That is human nature and, indeed, progress is often accelerated that way. However, such scientists are usually confronted with inhabitants that have grown up there and have thus acquired a superior knowledge of the terrain. The authors of this book belong to the second species. They enable us to participate in the knowledge accumulated in the realm of nonlinear dynamics, in particular nonlinear oscillations and waves, since about the turn of the century in the former Soviet Union. The political division of the human world created a separate community working on the foundations of nonlinear physics and hardly interacting with its counterparts elsewhere. Nevertheless, in the course of time, the most famous results crossed the divide and were incorporated into the common knowledge of mankind. The list of famous names is impressive: Landau (e.g. nonlinear Landau damping, Ginzburg-Landau equation), Kolmogorov (turbulence theory, Kolmogorov entropy), Oseledec (ergodic theorem), Lyapunov (Lyapunov exponents of dynamical systems), Mandelstam (wave scattering), Andronov (nonlinear oscillators), Arnold (Arnold tongues, Arnold diffusion), Melnikov (Melnikov criterion), Chirikov (resonance overlap) and many others. I apologize for not citing them all. The experience of all of them is incorporated here and will equip the reader with a firm understanding of the new "chaos theory" or rather "nonlinear physics" now under construction.

I read this book with great pleasure and benefit and am sure that every reader will do the same!

Rossdorf, March 1992 *Werner Lauterborn*

Preface

Dear reader! This book is neither a monograph nor a textbook. The intentions of the authors are to review past accomplishments and to give a few guideposts to future developments in that fascinating interdisciplinary area of modern physics, *nonlinear physics*. Keywords in this new field are *dynamical chaos*, *solitons*, *self-organization*, *turbulence* and *structures*. Nonlinear physics – or, more generally, the theory of nonlinear phenomena – emerged only in the last quarter of a century as an independent branch of science. It should be borne in mind that its rapid and distinct growth is largely due to the fact that it was constructed on the very solid and reliable foundations of nonlinear mechanics and the theory of nonlinear oscillations and waves. This provides the motivation for including in the title of this book "oscillations" alongside "chaos" and "structures".

Nowadays, nonlinear physics is so closely related to the theory of nonlinear dynamical systems that one often regards the terms "nonlinear physics" and "nonlinear dynamics" as synonymous.

The basic building blocks in the foundations of nonlinear sciences were laid by H. Poincaré and A.M. Lyapunov, and later by L.I. Mandelstam, A.A. Andronov, D. Birkhoff, B. van der Pol and others. The contributions by Mandelstam and Andronov distinguish themselves partly because of the importance of the specific results obtained (Mandelstam-Brillouin scattering, Andronov-Hopf bifurcation, etc.). Yet, their impact on this modern branch of science is more fundamental since they created a general attitude and approach to the problems of nonlinear systems, suggesting a "complete" investigation of them, including a study of all details and of the effects that varying parameters have on their evolution. They also constructed basic models and developed the "international" interdisciplinary language still used by the scientific community. It is for these reasons that we dedicate our book, which might also be entitled *Nonlinear Physics: Yesterday – Today – Tomorrow*, to L.I. Mandelstam and A.A. Andronov.

Nizhny Novgorod,
Summer 1992

A.V. Gaponov-Grekhov
M.I. Rabinovich

Contents

1. Introduction 1

2. Nonlinear Oscillations and Waves. Classical Results 11

2.1 Oscillators 11

2.1.1 A Marble in the Chute 11

2.1.2 Spring Pendulum and Nonlinear Optics . 14

2.1.3 Nonlinear Landau Damping and Amplification 18

2.2 Solitons 22

2.2.1 The Fermi-Pasta-Ulam Paradox 22

2.2.2 Solitons as Particles 25

2.2.3 Solitons and Shock Waves 27

2.3 Self-Excited Oscillations 31

2.3.1 Examples and Definitions 32

2.3.2 Competition and Synchronization 36

2.3.3 Self-Excited Oscillations in Chains and Continuous Systems 38

2.4 Bifurcations 40

2.4.1 Acquisition of a New Quality 40

2.4.2 Bifurcations of Equilibrium States 43

2.4.3 Bifurcations of Periodic Motion 45

2.4.4 Bifurcations – Changes of Stability in Periodic Motion . 45

2.5 Modulation 46

2.5.1 The Role of Small Parameters 46

2.5.2 Running Mandelstam Lattices. Modulation of Waves by Waves 48

2.5.3 Generation of Modulation 51

2.5.4 Self-Modulation 52

2.5.5 Recurrence 54

2.5.6 Modulation Solitons 56

3. Chaos 59

3.1 Historical Remarks 59

3.2 Marble in an Oscillating Chute 60

3.3 Stochastic Self-Excited Oscillations 64

3.3.1 The Lorenz Attractor 65

3.3.2 Synchronization – Beats – Chaos 68

3.3.3 Autonomous Noise Generator 69

3.3.4 Scenarios for the Birth of Strange Attractors 72
3.4 Chaos and Noise 75
3.4.1 Dimension and Entropy 75
3.4.2 The Cantor Structure of a Strange Attractor 76
3.4.3 Dimension and Lyapunov Exponent 78
3.4.4 Deterministically Generated and Random Signals 81

4. Structures 85
4.1 Order and Disorder – Examples 85
4.2 Attractors and Spatial Patterns 90
4.2.1 Examples of Equations 90
4.2.2 Multistability. Defects 92
4.3 Self-Structures 98
4.3.1 Convective Self-Structures 98
4.3.2 Localization Mechanisms 100
4.3.3 Self-Structures in Three-Dimensional Media 101
4.3.4 Interaction of "Elementary Particles" ... 103
4.3.5 Birth and Interaction of Spiral Waves ... 105
4.4 Attractors – Memory – Learning 107
4.4.1 How to Remember 107
4.4.2 "Camera + TV + Feedback" Analogue .. 109
4.4.3 Critical Phenomena 112
4.4.4 Structures in Neuron-Like Media 113

5. Turbulence 117
5.1 Prehistory 117
5.2 Basic Models of Dynamic Theory 120
5.3 Turbulence and Structures in Two-Dimensional Fields 122
5.3.1 Experiments 122
5.3.2 Development of Turbulence and Multi-Dimensional Attractors 125
5.4 Spatial Evolution of Turbulence 127
5.4.1 Flow Dimension 127
5.4.2 Spatial Bifurcations 129
5.5 Discussion 130

6. Nonlinear Physics – Chaos and Order 133
6.1 The Where and the How 133
6.2 Randomness Born out of Nonrandomness 134
6.3 An Unstable Path and Steady Motion. Are They Incompatible? 136
6.4 Does Chance Rule the World? 137
6.5 What is the Character of Nature? Integer or Fractal? 139

6.6 Fractal Fingers 141
6.7 Self-Organizing Structures 143
6.8 Singles 145
6.9 The New Life of an Old Problem 145
6.10 Spatial Evolution of Disorder 146
6.11 What Does Your Camera See When It is Watching TV? 147
6.12 Multistability and Memory 148
6.13 Nonlinear Dynamics in Society 149

Color Plates 151

Literature 177

Acknowledgements of the Figures 185

Subject Index 187

1. Introduction

L.I. Mandelstam
4.5.1879–27.6.1944
Born in Mogilev. University of Novorossisk (Odessa); graduated from Strassbourg University (1902) where he stayed up to 1914; chair physics at the Polytechnical Institute in Odessa; chair of the Institute of Theoretical Physics at Moscow University and in parallel work at institutes of the Academy of Sciences.

Work related to optics, quantum electronics, the theory of nonlinear oscillations, quantum mechanics, history and methodology of physics.

In the first steps towards the construction of a quantitative picture of the world, practically all branches of natural sciences (including physics, but with the possible exception of classical mechanics) limited themselves to the study of the most stable and most prominent fragments of this picture, that is, to equilibrium or quasi-equilibrium states and processes. But the further, natural progress of scientific development led inevitably to the need to solve also intrinsically dynamical problems. Examples are the transition of a system from one quasi-equilibrium state to another and the formation and decay of bound states of nuclei, atoms, stars or galaxies. The variety and sophistication of physical problems at the "dynamical" level appeared to be so complex that the usual *a priori* assumptions of weak fields and small perturbations were rather natural and more often than not practically the only possible approaches to unravel at least some of the features of quasi-equilibrium phenomena. This attitude characterizes the historical developments in optics, acoustics, electrodynamics and the majority of other branches of physics. Such an attitude was fostered by the fact that the experimenters investigating these phenomena had in most cases only rather weak fields at their disposal.

Consequently, the linear superposition principle – reflecting the hypothesis that the additivity of causes leads to the additivity of the resulting effects – seemed to be so natural that many people took it for the universal key to the understanding and quantitative description of most problems posed by nature. In solving problems not involving serious deviations from equilibrium states physicists continued to work within the realm of linear (or more exactly, linearized) dynamical models. In spite of the fact that even in those days the limitations of linear physics were quite obvious, linear approaches nevertheless remained the dominant ones for a long time. The respective mathematical concepts and tools like spectral representation, eigenfunctions of boundary value problems, etc. appeared so natural, simple and pictorial that they even started to acquire their own life independent of any applications. Relying on those methods it was often possible to understand the prominent features of the solutions of a problem without solving the related equations.

A.A. Andronov
11.9.1901–31.10.1952.
Born in Moscow. Graduated 1925 from Moscow University, since 1931 at Nizhny Novgorod (then Gorkii) University.

Work related to the theory of oscillations, general technical dynamics, theory of differential equations, control theory, quantum electronics, history of engineering.

But at the beginning of the twentieth century, the situation changed dramatically. The number of urgent nonlinear problems whose solutions could no longer be postponed, started to build up like an avalanche. In earlier times such problems were connected with traditional nonlinear mechanics (like the three-body problem and the description of waves on the surface of a fluid) while the pertinent nonlinear problems of the first three decades of our century arose largely in acoustics, solid state physics and statistical physics. Intrinsically nonlinear problems that emerged with the advent of radio engineering (like detection and generation of oscillations), arose as well in other applied problems (for example, in control theory). Nonlinear mathematics started to be not just a useful but also an indispensible tool for physicists. The fruitful interaction between the two disciplines led eventually to the birth of nonlinear physics. However, in those days the "nonlinear difficulties" seemed to be specific for each individual problem without being linked to each other.

It was only recognized in the nineteen twenties and thirties that deviations from the additivity of the responses to additive actions are not the exception but rather the rule. The same holds for the existence of systems in which the knowledge of an arbitrarily large number of individual solutions is still insufficient to predict the further motion of the system. In all of these cases the implications are that the linear approach is no longer applicable and it was realized that *non*linear problems are of fundamental importance for shaping our understanding of very different phenomena. It became clear that nonlinear problems from various branches of physics – and not only from physics – have very much in common and demand a unified approach for their description. Amongst the physicists working in rather different branches a "nonlinear way of thinking" developed and different parts of the sciences started to interact and to take advantage of the "nonlinear experience" gained by the others. It was in particular the work of *L.I. Mandelstam, B. van der Pol and A.A. Andronov* [1.1–3] that gave rise to the theory of nonlinear oscillations as the science describing nonlinear phenomena in different discrete systems [1.4].[1]

It was the experience gained within the theory of nonlinear oscillations which is at the foundation of the branch of sciences which we nowadays refer to as *nonlinear dynamics*.

J.W.S. Rayleigh
1842–1919. Awarded the 1904 Nobel Prize for physics for his investigations on the density of the more important gases and for his discovery of argon, one of the results of those investigations.

R.L. Weber
"Pioneers of Science"
(Hilger, Bristol, New York 1988)

In those early days it appeared to most physicists only natural to apply to these new nonlinear problems the powerful, formally well-founded and heuristically tested approaches which were (and are) so successful in the realm of linear physics. It is remarkable that it was indeed possible to progress along those lines sufficiently far to develop the linear methods to their perfection by treating the nonlinear effects as small corrections to or slowly varying parameters of the

[1] The need for such a theory was already felt by Rayleigh who published the first book on the theory of oscillations more than a hundred years ago. Although it is entitled *Theory of Sound* [1.5], the topics covered include electromagnetic oscillations as well as waves on the surface of fluids and oscillating flames. It is remarkable that so long ago, a scientist already stressed the fundamental importance of nonlinear problems. In particular, Rayleigh extracted the most prominent features of systems giving rise to oscillations. Andronov called them later "self-(excited)oscillations". Rayleigh was the first to present the nonlinear differential equation of auto-oscillations or self-excited oscillations. In a different context it was later on rediscovered by *van der Pol* [1.2].

Sir R.E. Peierls
A distinguished physicist (Oxford), important contributions to solid state physics, statistics, quantum mechanics and other branches of physics. As an exception to the pattern established for various reasons he collaborated with Soviet scientists not just once: in spite of 'hot' and 'cold' war he stayed in close contact with them over decades – a living bridge.

linear solutions assumed to represent the fundamental properties of the system. One of the first great advances along these lines was the work of *van der Pol* [1.2] who developed in the twenties the theory of quasi-harmonic oscillators in the radiofrequency range as generated by a triode. Another example is the work of R. Peierls who put forward the theory of heat transfer in crystalline media as being mainly due to anharmonic motion of the atoms in the lattice (see, e.g. [1.6]).

The historical developments related to our understanding of the nonlinear dynamics of systems with continuous parameters evolved in a similar way. Of course, again nonlinear processes like the evolution of shock waves [1.7] and water waves [1.8] were known a long time ago together with the fact that they do not allow for a quasi-linear description. However, practically up to the middle of the sixties these systems were not considered to be of particular interest and the specialists were largely engaged in work related to *weakly* nonlinear phenomena.[2]

The problems that can be interpreted in the language of weakly interacting linear objects (harmonic oscillations, waves, wavepackets, etc.) are the subject of *quasi-linear physics*. In the case of the by now classic nonlinear dynamics of discrete systems (e.g. in radio engineering and control systems [1.2, 9]) it was possible to obtain superb results within this framework. They include the key problems of nonlinear optics (the generation of optical harmonics [1.10, 11] and of coherent optical radiation in lasers [1.12, 13], the self-focussing of light [1.14], phase conjugation [1.15]). The same approach made it possible to interpret the nonlinear evolution of plasma instabilities (decay [1.16] and modulation [1.17] of an instability, nonlinear Landau damping and amplification [1.18]) and to develop the theory of weak turbulence in hydrodynamics [1.19] and in plasmas [1.20], etc.

The equations of quasi-linear physics are usually formulated for the parameters of weakly interacting elementary linear excitations or modes (like the amplitudes and phases of quasi-harmonic waves). The equations are constructed by modifying (locally in time and space) the parameters of the linear modes in such a way that the arising corrections (due to perturbation theory) to the approximate solutions will not make a significant contribution to the main part of the solution. Such procedures are successful because only corrections coinciding with one of the eigenmodes of the unperturbed problem will give rise to growing or resonance contributions (with small numbers in the denominator) to the main linear part of the solution. The summation of the resonances in the different orders of perturbation theory does not

[2] As a rule these phenomena were well described by the aid of traditional linear concepts like quasi-particles (photons, phonons, etc.) and harmonic wavepackets that vary gradually due to their self-interaction (or nonlinearity) or that act on each other through their potentials. Within such an approach only those processes appeared to be tractable for which the resonance conditions were fulfilled (or with energy and momentum conservation laws for the quasi-particles in each elementary interaction). For harmonic waves the synchronization condition $\omega_j = \sum \omega_i$ with $\mathbf{k}(\omega_j) = \sum \mathbf{k}_i$ corresponds to these requirements.

lead to a qualitative change of the main part of the solution, but only to small changes in the parameters representing its different modes (like amplitudes and phases of interacting waves). Similar summations lay at the foundations of nearly all methods of quasi-linear physics, e.g. the methods of *Krylov* and *Bogolyubov* [1.21], of Feynman diagrams [1.22], and others. Whenever it is impossible to account for all resonances, the quasi-linear approach fails and the researcher is confronted with a "strongly nonlinear" problem.

N.N. Bogolyubov
Born in 1909. Worked from 1928–73 at the Ukrainian Academy of Sciences (since 1965 as the director of the Institute of Theoretical Physics in Kiev), from 1936–59 also professor at Kiev University, in parallel since 1948 at the Steklov Institute for Mathematics of the USSR Academy of Sciences, since 1956 at the Joint Institute of Nuclear Research (Dubna; since 1965 as director), since 1950 also professor of Moscow University.
Work related to statistical, elementary particle, quantum and mathematical physics (e.g., nonlinear oscillations) and to quantum field theory.

In spite of the remarkable success of quasi-linear physics, it was not able to solve the strongly nonlinear problems that were of considerable interest to the researchers from the very birth of the theory of nonlinear oscillations onwards. The solution of these problems was facilitated by the development of the qualitative theory of dynamical systems by Poincaré and Andronov. However, up to the sixties, the general interest was focussed on the simplest behaviour of systems with a small number of degrees of freedom. (For continuous systems that concerned only particular forms of motion like stationary waves, see below.)

Speaking about the analogies between oscillations and waves, it has to be noted that they are very fundamental and manifold. It suffices to recall the well-known analogy between the spatial pulsations of interacting waves and the temporal pulsations of oscillations. Similarly far-reaching is the analogy between oscillations and waves possessing spatial structure and interacting in time. There are also more non-trivial analogies like the ones between nonstationary wave effects (e.g. periodic modulation waves) and the interaction processes of oscillations in an ensemble of coupled nonlinear oscillators (recurrence, quasi-periodicity, etc.). However, in the discussion of these or similar analogies arises the question: why and to what extent is it possible to compare a (continuous) wave system with a finite-dimensional one (more precisely, with a system possessing a small number of degrees of freedom)? In other words, under what conditions is it possible to reduce such a problem to the analysis of a low-dimensional phase space?

Even nowadays the answer to this question is not too obvious. At the beginning of the sixties some light was shed on this problem by the thorough investigation and analysis of nonlinear wave processes in the two extreme cases in which we have either media with strong dispersion and small nonlinearity or with strong nonlinearity and weak dispersion.

In the evolution of a wave, say, in a compressible gas or on the surface of shallow water (with no dispersion) the crest of the wave moves faster than its base. The nonlinearity distorts the wave continuously so that its profile becomes eventually ambiguous and breaks. That happens even for small nonlinearites and waves of arbitrary finite amplitudes. Obviously such a process cannot be expected to be describable within a finite-dimensional model. These circumstances may most conveniently be explained in the language of the rather pictorial spectral representation. In a medium without dispersion, the phase velocities of small perturbations are known to be the same for arbitrary frequencies. That explains why even weak harmonics as generated by nonlinearities, are in resonance with the "basic" wave (synchronization) and effectively excite it. Thus it is appreciated that any attempt to model such processes with the aid of a set of harmonics would make it necessary to include an infinity of them.

However, for weak nonlinearity and large dispersion of the system (as in the case of materials used in nonlinear optics) its synchronization – i.e. temporal and spatial resonance – may imply that only a few waves or frequencies contribute significantly so that the direct analogies with processes in oscillating systems with a small number of degrees of freedom are applicable.

It is remarkable that these modern interpretations of the effect of dispersion (for bounded systems with nonequidistant spectra) on processes in continuous nonlinear systems were already given by Mandelstam and his collaborators in the middle of the thirties. Admittedly they were concerned with bounded continuous systems with discrete nonlinearities. However, in the present context that is not an important distinction. This point is discussed in some detail in their contribution *New Investigation of Nonlinear Systems* to the Congress of the International Radio Union in 1935. With regard to systems with strong dispersion in which the distribution of the *overtones is nonharmonic*, it is stated that *in this case the stationary oscillations may have nearly sinusoidal shape. By the aid of an approach analogous to the theory of small parameters* (perturbation theory) *for systems with a finite number of degrees of freedom, it is possible to evaluate the amplitude, solve the stability problem.* In the other limiting case the dispersion vanishes, i.e. one has a *harmonic distribution of the overtones* (this is the problem which A.A. Vitt investigated in the context of the analysis of excitations of the strings of a violin by the bow) and *stationary oscillations are always strongly nonsinusoidal.*

The violin is about 60 cm in overall length, with its four strings tuned to G_3, D_4, A_4 and E_5.

Its relatives also have four strings: the viola, about 75 cm long, is tuned a fifth lower to C_3, G_3, D_4 and A_4; the cello (more correctly the violoncello), about 120 cm long, again an octave lower, tuned to C_2, G_2, D_3 and A_3; the double bass (also called the contrabass), around 200 cm in length, its strings tuned (this time in fourths) to E_1, A_1, D_3 and G_2.

I. Johnston
"Measured Tones"
(Hilger, Bristol, New York 1989)
p. 137

It is already 30 years since nonlinear problems were posed and solved in the general theory of relativity, in unified field theory, in astrophysics and naturally as well in hydrodynamics. The creation and annihilation of elementary particles, the evolution of the Universe, the emergence of random behaviour in simple systems without noise and fluctuations, the transition of the laminar flow of a fluid into turbulence and the generation of sophisticated spatial structures did not fit into the picture of quasi-linear physics. In the course of time the pertinent nonlinear problems were steadily increasing in number and significance. This explosion is particularly obvious in our days.

Let us stress that intrinsically nonlinear problems are already a substantial part of quasi-linear physics itself. Having exploited the quasi-linearity of a problem, we obtain a new simpler dynamical model (for amplitudes and phases of the oscillations) which in general no longer contains small parameters. The investigation of such models outside the realm of equilibrium states is a problem of nonlinear physics. Naturally, it is possible that within the frame of the new approximate model there may as well be small parameters so that the procedure

Nonlinearities comprise both stochastic, turbulent behaviour and the generation of deterministic structures.

may be repeated. However, in that case the encounter with the intrinsically nonlinear problem is only once more shifted to the next level of sophistication in the description.

In this book we will more closely inspect the two – to our understanding – most general and fundamental nonlinear problems: the birth of stochastic, turbulent behaviour of dynamic systems (unaffected by fluctuations) and the generation of deterministic structures in nonlinear media (which occurs often in spite of fluctuations). These problems which are related to weather forcasting, the heating of plasmas, biophysics, unified field theory, evolution theory etc. are nowadays not just in the focus of the interest of physicists but concern mathematicians too. It was particularly the mathematicians who made a lot of progress in the last decade providing the more applied physicists with new exact results and handy tools for the investigation of such systems. The most prominent ones amongst them are the (particle-like) *soliton* and the *strange attractor* or, in a more general sense, the understanding that there exist definite connections between the instability of motion and the birth of chaos.

Nowadays the problems of dynamical chaos and pattern formation are some of the hottest topics in nonlinear physics. Looking backwards we may say that after the re-birth of the soliton, the discovery of strange attractors and stochastic behaviour of dynamical systems, the observation of the universality of mechanisms and models of self-organization in systems and media of different nature, there emerged the new interdisciplinary branch of science called *nonlinear dynamics* or the theory of strongly nonlinear phenomena.[3]

After all, what do – at first sight very different – phenomena like the transition of a laminar flow to turbulence, self-organization, solitons have in common? The most important point is the fact that all these phenomena require in a certain sense a global description of the system, i.e. a description at large, which does not allow the restriction of the analysis of the behaviour of the nonlinear system to neighbourhoods of individual equilibrium (i.e. linear) states. Let us discuss this in more detail.

A well-known and possibly the simplest example for self-organization is the formation of a stationary progressing front of flames: the

---

[3] At the beginning of the seventies it appeared as if the situation known from the early thirties was repeating itself: as in previous times, mathematical developments provided the foundations for new branches of nonlinear science. However, before the time was ripe, they did not acquire the status of working tools for physicists, chemists, biologists, and others investigating the manifold of phenomena offered by Nature. The importance of these new discoveries and the need to develop them still further was only realized by a small circle of professional mathematicians. And it was only in the middle of the seventies when a huge army of theoreticians and experimentors was confronted with the generality of the phenomena of chaos and pattern formation that the ice broke and an intensive exchange of ideas across the borders of the different branches of nonlinear sciences took place. In that situation pure mathematics received a strong impulse to take over from nonlinear science, the formulation of new problems, and again this topic became the object of pagan idolatry among the applied sciences.

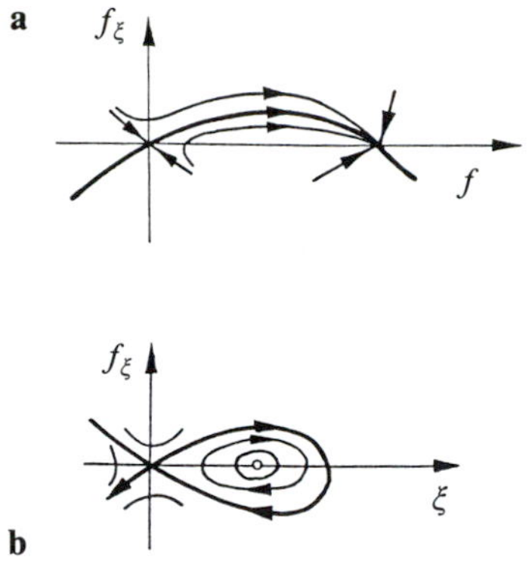

Fig. 1.1a,b. Specific solutions of partial differential equations in phase space; the separatrix is represented by the *bold* curve.
a) Stationary shock wave in a dissipative medium.
b) Separatrix as the image of a solitary wave in a conservative medium

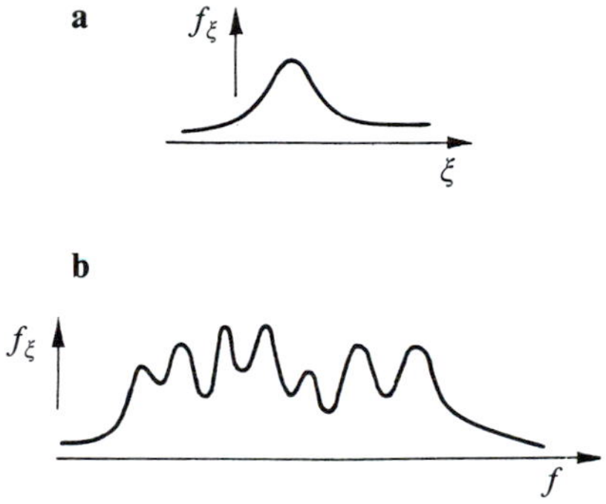

Fig. 1.2a,b. Solitary wave:
a) without structure and
b) with internal structure

burning process itself takes place in a relatively thin region separating the cold burning from the products of the burning; this region in which oxydation occurs moves relative to the hot matter with constant speed independent of the initial conditions [1.23]. This one-dimensional wave front of burning may be modelled by the partial stationary wave solution $f(\xi) = f(x - vt)$ of a system of differential equations. In phase space this solution is represented by a particular trajectory: the separatrix connecting two equilibrium states (Fig. 1.1) one of which corresponds to the values before the front ($\xi \to \infty$ where the reaction has not yet been initiated) and the other to the ones behind the front ($\xi \to -\infty$ where the reaction has been completed). Similar nonlinear solutions correspond to stationary solitary waves.

In conservative media they are represented by solitons. The separatrix reflecting the solitary wave does not go over into another equilibrium state but returns to the initial one (Fig. 1.2). We have a separatrix loop. When the stationary waves are described by equations of an order higher than two, the solitary wave may have an extremely complicated internal structure (see Fig. 1.2b).

Thus even for the description of stationary waves – like solitons and fronts of evolving flames (and the structure of the front of shock waves) – it is essential to have a solution of the nonlinear problem which describes the complete transition process from one equilibrium state into another one and thus also the changes of the variables (say, $\xi = x - vt$) in an infinite interval. This implies a solution which in a certain sense is, at the same time, well localized and valid for infinite values of space and time coordinates. These particularities are what we had in mind when speaking about the need to obtain a global description for such systems.

The connection of stochastic dynamics with a global description, with the behaviour of the system at large, which leads to the necessity to investigate the general structure of the evolution of the system is even more fundamental and requires a detailed discussion. This is due to the fact that we have in this context to renounce well established and seemingly most natural physical concepts.

Indeed, in line with the traditional way of thinking we allow for the possibility of random behaviour of a quasi-linear system like an ensemble containing a large number of weakly interacting "almost linear" oscillators. This assumption reflects the implicit hypothesis of a chaotic behaviour of the phases of the individual oscillators which allows us to go over to an averaged description employing the intensities of the oscillators or the numbers of quanta as useful quantities. An example is the theory of weak turbulence (e.g. [1.19, 20]). In such an approach there is indeed no need to consider a strong nonlinearity, neither is there intrinsic dynamical stochastization. The mechanism for randomness is not explained but simply hidden in the initial hypothesis.

Stochastic behaviour of dynamical systems is by no means a consequence of the overlapping of a large number of independent motions. Neither is this phenomenon related to the popular notion of stochasticity which is viewed as an approximation to something that cannot be described exactly, to something "intrinsically random". Even now many physicists consider the word combination "dynamical stochastization" to be an intrinsic contradiction. This is due to the fact that the consensus is that dynamics refers to a completely determined evolution process of a physical system whose past, present and future are unambiguously programmed by the original equations and the chosen inital conditions. If the system is a complicated one (implying that it contains a large number of sub-systems) then a statistical approach may still be sensible as a means of describing complicated, yet, intrinsically non-random motion. However, for a simple system like two interacting nonlinear oscillators the basic question is: if there exists any randomness at all where could it come from? – The answer is unexpected: *first*, randomness in such a system is possible and *second*, even when probing it from different points of view such a motion cannot be distinguished from "truly" random motion (for example, under the influence of external random forces) and is a particular case of deterministic strongly nonlinear systems! In terms of physics that means the following: although each trajectory of the system is deterministic in the above sense, all (or almost all) finite trajectories are unstable "at large". After a sufficiently long time this leads to a loss of information on the past of the system (i.e. on its initial conditions) independent of the (finite) accuracy with which that past had been specified initially.

H. Poincaré (1854–1912) remarked on the nature of accidental events, of chance:

Thus, stochasticity – that is the chaotic motion of a dynamical system – is the result of the instability of trajectories located within a limited yet sufficiently large region of phase space (and not just in the vicinity of equilibrium states). The proof for the existence of such trajectories and the prediction of the statistical properties of their corresponding motions demand an investigation of the general structure of the motion of the nonlinear system at large. That is the source for the true nonlinearity of the problem.

In connection with such problems demanding the study of the behaviour of systems at large, the inherently nonlinear language infiltrates deeper and deeper into physics and the quantitative theory of differential equations and the theory of dynamical systems. Back in the thirties this process was initiated when A.A. Andronov related the periodic self-excited oscillations in nonlinear nonequilibrium systems to Poincaré limit cycles.[4] Nowadays associated concepts like,

[4] Quite a few of the discoveries in the theories of nonlinear oscillations and dynamics had their lonely pioneers for which time was not yet ripe. As mentioned previously, the need to study particular problems within physics, say, may not just stimulate new mathematical discoveries, but also bring back to life almost forgotten results and methods. That held true for limit cycles which were already discovered before the time of Poincaré: *To avoid twisting the historical perspective, it is necessary to make the following remark. Ten years before the discovery of the radio, the French engineer Leaute (1885) studied self-oscillations in an automatic control mechanism, investigated the phase space of that system and plotted its integral curves and limit cycles (without giving them those names: apparently he was not acquainted with the work of Poincaré published a bit earlier and in which limit cycles appeared for the first time within the context of mathematics)* [1.24].

e.g. separatrix, periodic trajectory, Poincaré mapping and others are for physicists associated with real phenomena encountered in nature. The related formalism became gradually a powerful instrument in the intuitive analysis in nonlinear physics – similar to spectral analysis in linear physics. At present the study of nonlinear physics is characterized by a combination of strong analytical methods and of qualitative theory and numerical experiments.

... quite negligible reason, escaping our attention due to its smallness, can cause a considerable effect, which we cannot foresee, and in such a case we say that the phenomenon is the result of chance.

Coming to the end of this introduction, we would like to give a short formulation of the "Mandelstamian" approach to nonlinear phenomena: *On the one hand, the ability to understand from a unified position the manifold of rather different phenomena, with a limiting precision observe their common properties and combine all prominent features into a pictorial model; on the other hand, a strong interest in the particular individuality of a physical phenomenon.* In this book we attempt to present current nonlinear physics from the point of view of Mandelstam and Andronov. The architecture of the treatment is as follows.

Attempting to keep the historical perspective, we consider in Chap. 2 with the title *Nonlinear Oscillations and Waves - Classical Results* from the current point of view the classical results of the nonlinear theory of oscillations and waves. In Chap. 3, *Chaos*, we present our perspective of the most prominent results about the stochastic behaviour of deterministic finite-dimensional systems. The problems of self-organization and the emergence of specific shapes is the subject of Chap. 4, *Structures*, which is based on a unified approach to phenomena of rather different nature. Chap. 5, *Turbulence*, is based on the material elaborated on in the first chapters and is concerned with the analysis of chaotic dynamics of nonlinear fields, in the first place by making reference to turbulence of the flow of fluids.

Of course, the task of giving readily accessible and exact information on the most prominent features of nonlinear physics, can only be completed if use is made of the classical results of the nonlinear theory of oscillations. In doing so, we were forced to suppress several very important notions inclusion of which would lead us too far astray (e.g. the renormalization group method for turbulence, the theory of fractal structures, etc.). However, we hope that even in the rather incomplete and personal presentation of this book, the reader will appreciate the depth and brightness of the ideas and results of modern nonlinear physics and that he will get a feel for its solid and manifold connection with the nonlinear physics of the days of Mandelstam and Andronov.

To draw attention to the beauty (even in the colloquial sense) of the subject and to highlight some of the problems of deterministic chaos, pattern formation and turbulence, we prepared for you the completely separate nicely illustrated part *Nonlinear Physics – Chaos and Order*, Chap. 6. It introduces not just some easily accessable basic notions, but exhibits also the beauty of the subject so that you may like to go first through this self-contained part of the book before continuing with the next chapter.

2. Nonlinear Oscillations and Waves. Classical Results

It has long been understood that real life and physics are actually governed by nonlinear laws. The most fascinating pages of nonlinear physics today are connected with keywords such as *chaos, structures* and *turbulence*. However, the now almost classical concepts of the rapidly evolving branch of nonlinear physics like *solitons, self-excited oscillations, shock waves* and others have not lost their significance and attraction. In this chapter we will discuss associated classical results. We do not have the space to provide a comprehensive treatment, thus we will often simply "show" the equations and say a few words about their features of interest in the given context. Our aim is to raise your appetite – not to satisfy it.

2.1 Oscillators

The simplest examples of oscillators are mechanical pendulum and swing. Below we shall consider idealized systems of relevance to subjects ranging from classical mechanics to optics and treat also realistic effects like damping and amplification.

2.1.1 A Marble in the Chute

A marble sliding in a bent chute is a pictorial and simple version of the nonlinear oscillator; an electrical circuit is a more "abstract" example, see Fig. 2.1. Because of its simplicity the harmonic oscillator has been thoroughly studied in the nonlinear theory of oscillations and in nonlinear dynamics. The general case of the nonlinear oscillator is described by

$$\frac{d^2u}{dt^2} = -\partial_u U(u,t) + g(\dot{u},t) \quad \text{with} \quad \dot{u} \equiv \frac{du}{dt} \quad \text{and} \quad \partial_u \equiv \frac{\partial}{\partial u}.$$

We may consider t to be the time and u the spatial coordinate (depending on the specific application, it may also stand for the electric current in a circuit, the density of chemically interacting substances etc.). $U(u,t)$ is the potential energy of the system and $g(\dot{u},t)$ describes dissipative effects. In the simplest case U and g are time independent and we have an autonomous oscillator. For $g(\dot{u}) \equiv 0$ the oscillator is conservative. Such systems are referred to as nonlinear oscillators in the narrow sense.

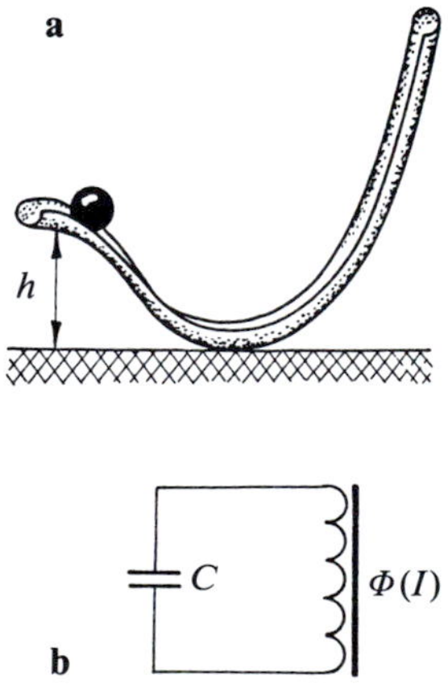

Fig. 2.1a,b. Examples of nonlinear oscillators:
a) marble in a chute;
b) electrical circuit

In most instances the nonlinear oscillator is the prototype of a dynamic system, that is, of a system whose time evolution depends for all times, including infinity, uniquely on the initial state.[1] To facilitate the subsequent discussion, let us recall some of the elementary concepts of nonlinear dynamics. One notion is the *state* of a system which is for a given moment of time specified by a complete set of variables (there may be different criteria for choosing a particular set of variables: symmetry considerations, physical origin, etc.). A set of states of the dynamic system forms a *phase space*. Each state corresponds to a point in this phase space and the evolution of the system is depicted by (phase) *trajectories*. The ensemble of states at a fixed moment of time is the *phase volume*.

The qualitative evolution of dynamic systems is read off the behaviour of the phase trajectories. Examples are:
a degenerate trajectory corresponding to a point in phase space depicts a state of rest (equilibrium); a closed trajectory represents periodic fluctuations; the trajectory of quasi-periodic motion with m incommensurate frequencies is an open winding on an m-dimensional torus; trajectories within the same part of phase-space (or "being dense in a certain subset of phase space" as a mathematician would put it) characterize a stationary regime of oscillations, the established motion of the system; trajectories that never come back to the vicinity of their starting points describe a regime of transient oscillations.

In his lectures on oscillations (1930–1932) *L.I. Mandelstam* [2.1] considered a nonlinear electrical circuit and the marble in a chute (Fig. 2.1) as the first examples of nonlinear oscillators. He remarked that it is reasonable to *imagine the entire qualitative picture of the motion on the basis of the differential equation itself without solving it.* For the simple nonlinear oscillator as a conservative nonlinear system with one degree of freedom this qualitative picture is read off its phase portrait, see Fig. 2.2.[2] The motion of the nonlinear oscillator is fully determined by its initial energy. At low energies it performs small, harmonic oscillations. As the energy is increased, the oscillations deviate more and more from the harmonic regime: most of the time the motion will take place in "slow" sections in which the marble rolls up to the top of the hump (Fig. 2.1a) and, eventually, at initial energies exceeding $E_0 = mgh$ the motion of the marble will no longer be pe-

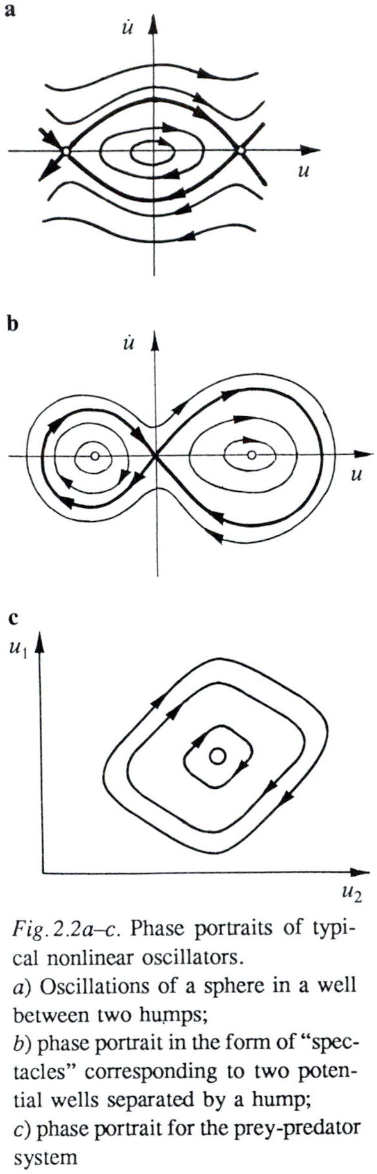

Fig. 2.2a–c. Phase portraits of typical nonlinear oscillators.
a) Oscillations of a sphere in a well between two humps;
b) phase portrait in the form of "spectacles" corresponding to two potential wells separated by a hump;
c) phase portrait for the prey-predator system

[1] Note that the system depicted by $\dot{u} = u^2$ (with the solution $u = u_0/(1 - u_0 t)$ which goes to infinity for the finite time $t_\infty = 1/u_0$, the so-called explosive instability, [M.I. Rabinovich, D.I. Trubetskov: *Introduction to the Theory of Nonlinear Oscillations and Waves* (Kluwer, Amsterdam 1989)]) does not belong to this class of dynamical systems. Similarly, the wave equation of the type $\partial_t u + a(u)\partial_x u = 0$ does not describe such a system, because the wave profile develops within finite time a step or shock so that the wave breaks, implying that its solution is no longer unique, see Sect. 2.2.3 *Solitons and Shock Waves*. However, nonlinear physics is also interested in such models with their particulars.

[2] Mandelstam stated [2.1]: *To have the appropriate feel for the necessary mathematical rigour is the most difficult problem for a physicist. It would be better to say that it is essential that he should to determine the appropriate measure of rigour.*

Analogies are useful for analysis in unexplored fields. By means of analogies an unfamiliar system may be compared with a better known system. The relations and actions are more easily visualized, the mathematics more readily applied, and the analytical solutions more readily obtained in the familiar system.

Harry F. Olson
"Dynamical Analogies"
(Van Nostrand, Princeton 1943)

riodic. In the phase plane this motion is represented by the separatrix passing from one saddle point to the other (Fig. 2.2b). Thus the motion of the nonlinear oscillator is nonisochronous implying that frequency ω and period T depend on the amplitude or energy. For motions that are not too close to the separatrix we can take the relations $\omega = \omega(A^2)$ and $T = T(A^2)$ to hold.

It is by no means a simple task to show that a given dynamic system whose phase space is in a plane belongs to the class of nonlinear oscillators, i.e. to establish that it is conservative. A rather easy case is

$$\ddot{u} - u(1 - u/2) = 0 , \tag{2.1}$$

whose phase plane is shown in Fig. 2.2b. Indeed, the integral of (2.1) is readily recognized as the energy integral $\dot{u}^2 - u^2 + u^3/3 = \text{const}$, while the system

$$\dot{u}_1 = u_1(\nu_1 - \varrho_1 u_2) , \quad \dot{u}_2 = -u_2(\nu_2 - \varrho_2 u_1) \tag{2.2}$$

appears at first sight to be nonconservative. (It describes the ecological problem of the interaction between two biological species, herbivores and carnivores.) The integral $\varrho_2 u_1 + \varrho_1 u_2 - \nu_2 \ln u_1 - \nu_1 \ln u_2 = \text{const}$ obtained by *Gauze and Vitt* [2.2], looks sufficiently complicated. The phase portrait of this nonlinear oscillator is shown in Fig. 2.2c. Such camouflaged oscillators are fairly often encountered in realistic problems.

Looking at the phase portrait of a nonlinear oscillator we can actually predict its behaviour for arbitrary situations, that is for arbitrary initial conditions. It may be used to model rather different phenomena. The most important cases correspond to (nearly sinusoidal) oscillations in the neighbourhood of stable equilibrium and to large amplitude oscillations in which the marble in Fig. 2.1 rolls up to the top of the hump (this is a particular manifestation of the nonisochronous behaviour of the nonlinear oscillator to be discussed below), the solitary oscillations corresponding to the separatrix loop that drops down to zero as $t \to +\infty$ and $t \to -\infty$. In the theory of dynamic systems the latter is known as a doubly asymptotic trajectory.

In the linear theory the analysis of the harmonic oscillator as the first step is followed by the investigation of coupled oscillators. Within nonlinear problems the next step towards more sophisticated systems is also related to systems with a larger number of degrees of freedom, that is to the analysis of two interacting nonlinear oscillators.[3] This problem will be considered below.

[3] It should once more be emphasized that such an "obvious" extension of a system is not necessarily unambiguously specified. As stated previously and to be repeated in the chapter on chaos, complex behaviour does by no means always imply complex systems. Thus the increase in complexity may be related to a change in the interaction of the system rather than to an increase in the number of its sub-units.

2.1.2 The Spring Pendulum and Nonlinear Optics

In 1931 there appeared a paper [2.3] in which *Fermi* reported on the way in which internal resonances of the CO_2 molecule give rise to their Raman spectra. L.I. Mandelstam suggested to *A.A. Vitt* and *G.S. Gorelik* [2.4] investigation of resonant-interaction effects in terms of nonlinearly coupled oscillators taking the spring pendulum without friction as a model, see Fig. 2.3a. The corresponding equations have the form

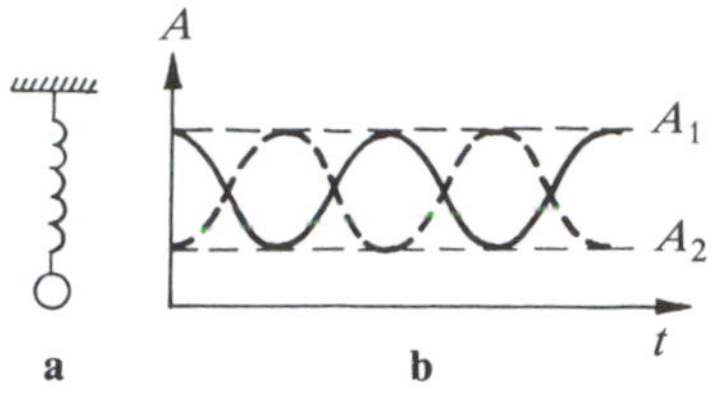

Fig. 2.3a,b. Oscillators:
a) spring pendulum;
b) periodic energy exchange between angular and vertical oscillations

$$\ddot{u}_1 + \frac{k}{m} u_1 = l \left(\dot{u}_2^2 - \frac{g}{2l} u_2^2 \right) , \tag{2.3}$$

$$\ddot{u}_2 + \frac{g}{l} u_2 = -\frac{1}{l} \left(\frac{g}{l} u_1 u_2 + 2 \dot{u}_1 \dot{u}_2 \right) . \tag{2.4}$$

This system of coupled differential equations was solved using the averaging method. It was established that at the parameter ratio $k/m \approx 4g/l$ with $\omega_{\text{vert}} \approx 2\omega_{\text{ang}}$, the energy is periodically pumped from angular (ang) to vertical (vert) oscillations and vice versa; a finding which was also confirmed experimentally [2.4]. Such energy exchange effects between different types of oscillations are already familiar from quasi-linear physics. Assuming a weak nonlinearity the averaged equations for amplitudes and phases of the ω- and 2ω-oscillators interacting in time (or space) may be written in the form [2.5]

$$\begin{aligned} \dot{A}_1 &= -\sigma_1 A_1 A_2 \sin \Phi , \\ \dot{A}_2 &= \sigma_2 A_1^2 \sin \Phi , \\ \dot{\Phi} &= -\left(2\sigma_1 A_2 - \sigma_2 A_1^2 / A_2 \right) \cos \Phi - \delta . \end{aligned} \tag{2.5}$$

In this expression we used $\Phi = 2\varphi_1 - \varphi_2 - \delta t$ and took δ to denote the detuning from exact resonance. With the help of the energy integral

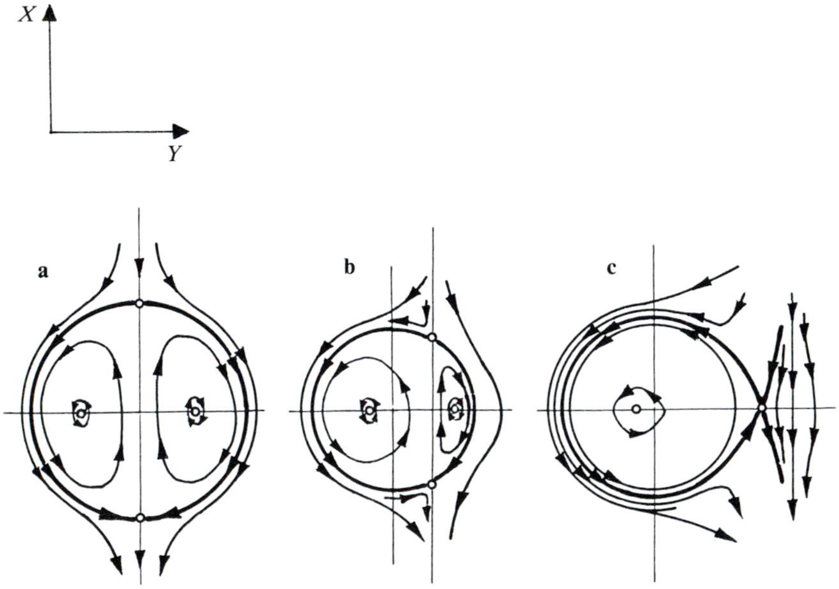

Fig. 2.4a–c. Phase portrait of a nonlinear oscillator describing the energy exchange between the two harmonics of a system with quadratic nonlinearity and detuning δ:
a) $\delta = 0$;
b) $|\delta|/2\sigma_1 A_0 < 1$;
c) $|\delta|/2\sigma_1 A_0 > 1$

$\sigma_2 A_1^2(t) + \sigma_1 A_2^2(t) = \text{const} = \sigma_2 A_1^2(0) + \sigma_1 A_2^2(0)$ and by introducing the new variables $X = A_2 \sin\phi$ and $Y = A_2 \cos\phi$ these equations are readily converted into the ones of a nonlinear oscillator. For various values of the detuning Fig. 2.4 shows the phase portraits of the resulting oscillator. Already at this rather low level of sophistication we have to deal with a strongly nonlinear system! Assuming only a weak nonlinearity or equivalently, weak initial excitation energies, a system of two nonlinearly coupled oscillators exhibits only very simple quasi-periodic motions. As far as the physics is concerned, the difference between various motions of this kind (see Fig. 2.3) reflects merely different energies and periods of the beats of these oscillators. We shall see that this simple feature is also inherent to many other nonlinear systems which at first sight appear to be very complex.

The next step in our considerations exploring more and more complicated systems would lead us to the study of weakly nonlinear interactions between three resonantly coupled oscillators and eventually to the same conclusions.

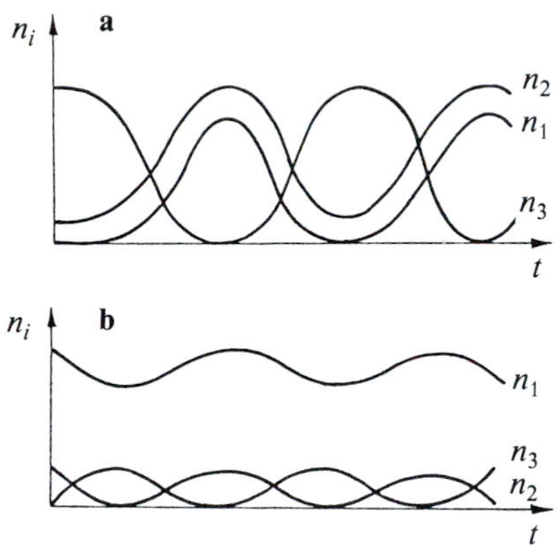

Fig. 2.5a,b. Periodic energy exchange in a system of three coupled nonlinear oscillators with frequencies $\omega_3 = \omega_1 + \omega_2$:
a) $n_3(0) \gg n_2(0)$ and $n_1(0) = 0$ and $n_2(0) = 0$;
b) $n_1(0) \gg n_3(0)$

The frequencies of these oscillators are related through $\omega_1 + \omega_2 = \omega_3$. In the case of quadratic nonlinear equations with slowly varying amplitudes $a_j(t)\exp[i\omega_j t]$ the oscillations can be represented by[4]

$$\dot{a}_1 = a_3 a_2^* \ , \quad \dot{a}_2 = a_3 a_1^* \ , \quad \dot{a}_3 = -a_1 a_2 \ . \tag{2.6}$$

Simple algebraic operations lead us from these equations to the *Manley-Rowe relations*

$$n_1(t) - n_2(t) = \text{const} \ , \quad n_3(t) + n_{1,2}(t) = \text{const} \tag{2.7}$$

for the oscillation intensities $n_j = |a_j|^2$. Analyzing these relations the characteristics of the interaction between the oscillators are readily appreciated. For example, if initially the energy of the low-frequency oscillators with ω_1 and ω_2 was small and the one of the high-frequency oscillator with ω_3 large, then the energy of the latter will effectively be taken over by the low frequency oscillators, see Fig. 2.5a. However, if at $t = 0$ one of the low-frequency oscillators had a larger portion of the initial energy then there would be almost no energy exchange, see Fig. 2.5b.

This characteristic of the interaction may be viewed as an analogue to decay and fusion processes of quantum mechanical particles or quasi-particles like phonons or photons. Naturally, quasi-particles can only fuse if there is something they can fuse with. Consequently the number n_3 of quanta with ω_3 born in such a fusion, will be exactly equal to the smaller of the quantum numbers n_1 and n_2 that existed in the prenatal period. The difference $n_1(0) - n_2(0)$ will, however, not be used up and be conserved for all times, $n_1(t) - n_2(t) = \text{const}$.

[4] The signs for the coefficients at the right-hand-sides of these equations follow directly from energy conservation that holds for this conservative system of coupled oscillators with:

$$\frac{d}{dt}\left(|a_1|^2\omega_1 + |a_2|^2\omega_2 + |a_3|^2\omega_3\right) = (a_1^* a_2^* a_3 + \text{c.c.})(\omega_1 + \omega_2 - \omega_3) \equiv 0 \ .$$

Obviously, the sums of the numbers of quanta that have been created n_3 and of those that have not yet been used up n_2 by the time t must also be conserved $n_3(t) + n_2(t) = \text{const}$. It should be noted that the constancy of the number of quanta of quantum mechanical oscillators implies that the number of quanta is an adiabatic invariant [2.6]. The adiabatic invariance is violated if the oscillator performs a transition from one level to another one which may be related to the resonant absorption of the energy of an external field with frequency Ω by the oscillator. Under suitable conditions, such a violation of adiabatic invariance can even occur for large degeneracies of the resonance with $\omega = m\Omega$ where $m \gg 1$, that is when the external field is varying very slowly. With regard to the classical oscillator, this result about the violation of the adiabatic invariant was first obtained by Mandelstam and his students *Andronov* and Leontovich back in 1928 [2.7].

For a different problem R.V. Khokhlov derived in 1961 equations almost identical to (2.5). He was considering the stationary interaction in space of a wave having the frequency ω_1 and the wave number k_1 with a second harmonic of frequency $\approx 2\omega_1$ and wave number $\approx 2k_1$. The only difference between the two approaches was in the derivatives: in the spatial problem the "slow" amplitudes and phases varied along the direction of propagation x rather than in time. It was found that in the propagation along x the wave with $2\omega_1$ first amplifies the initially (at the boundary) weak signal wave with ω_1 to transfer almost all the energy to it [2.5]. Subsequently, the reverse process occurs: the now strong ω_1-wave generates a second harmonic $\omega_2 = 2\omega_1$ and then the whole process starts anew. In other words, we observe exactly the same phenomenon of periodic energy exchange between harmonics as known from the spring pendulum (except for the fact that it is now in space rather than in time), see Fig. 2.3.

We shall see how sounds [waves] quarrel, fight, and when they are of equal strength destroy each one another, and give place to silence [darkness].

Sir Robert Stawell Ball "Wonders of Acoustics, or the Phenomena of Sound" (Scribner, Armstrong and Co., New York 1872)

It is remarkable that in the same year, 1961, the generation of second harmonics was observed in the propagation of a light wave from a ruby laser in a transparent nonlinear crystal, *Franken* [2.8]. Khokhlov's work and these experiments of Franken mark the beginning of the era of modern nonlinear optics.

Thus it is in decays of the type (2.6) as well as in subharmonic generation that the response to the pumping frequency ω_3 becomes significant as the amplitudes of the amplified waves (oscillators) with ω_1 and ω_2 increase and the decay process gives rise to fusion. In the next cycle the whole history repeats itself; for waves with a specified spatial structure in time or for stationary harmonic waves in space. In this way a system of three weakly nonlinear oscillators exhibits just simple periodic (or quasi-periodic) motion.

On the basis of the considered examples within the frame of an averaged description it could be inferred that a system of two (and even three) coupled oscillators is very simple in the sense that it exhibits no unforeseen behaviour. However, we shall not jump to conclusions, but instead have a look at a system of two nonlinear coupled oscillators with

$$\ddot{u}_1 + u_1 = -\mu 2 u_1 u_2 \ , \quad \ddot{u}_2 + u_2(1 - \mu u_2) = -\mu u_1^2 \ , \qquad (2.8)$$

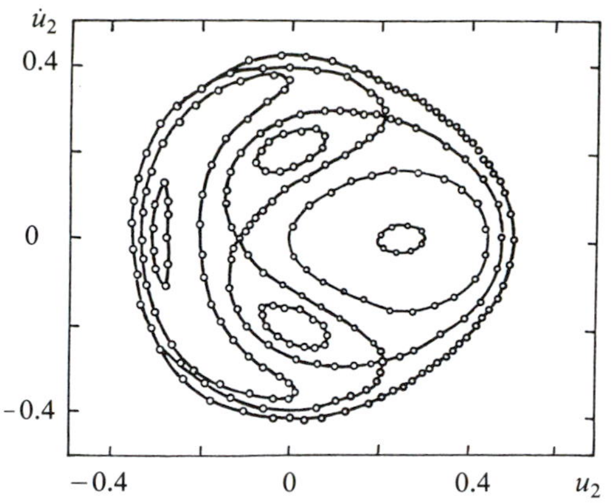

Fig. 2.6. Traces of a trajectory on a secant plane with $u_1 = 0$ in the phase space of the system (2.8) at $\mu = 1$ and at the initial energy $E_0 < 1/12$, see [2.9]

to understand their behaviour outside the quasi-linear regime. The equations are as simple as the ones of (2.4). In these cases the averaging method fails, but numerical experiments show[5] that for small μ with $\mu \ll 1$ the oscillators exhibit a simple quasi-periodic motion. This holds also for μ being of the order of unity and small inital excitation energies. Figure 2.6 displays the corresponding intersection of the plane $u_1 = 0$ with a trajectory in three-dimensional $(u_1, u_2, \dot{u}_2)$ phase space (2.8). This space becomes three-dimensional if for $\mu = 1$ the energy integral [2.9]

$$E = \left(\frac{1}{2} + u_2\right) u_1^2 + \left(\frac{1}{2} - \frac{u_2}{3}\right) u_2^2 + \frac{1}{2}\left(\dot{u}_1^2 + \dot{u}_2^2\right) \tag{2.9}$$

is taken into account. It looks as if all trajectories lay on smooth surfaces (tori), that is the motion of the system is conditionally periodic for arbitrary initial conditions. After all, system (2.8) is apparently as simple as expected at first glance! But let us have a look to see what happens if we increase the oscillation energy. First of all, the motion of the second oscillator will become strongly nonlinear. Motions close to the separatrix of a single nonlinear oscillator will appear (see Fig. 2.2b) and because of the presence of the "external" force $u_1^2(t)$ we can no longer say whether they will remain quasi-periodic or whether the type of motion will change and go over from the region within the separatrix to the exterior region.

Figure 2.7 shows the results of numerical experiments performed with two coupled nonlinear oscillators (2.8) with initial energies $E_0 > 1/12$. For initial energies exceeding $E_0 = 1/12$ by only 0.04, yet still corresponding to simple motions, the phase trajectory no longer winds around a certain surface, but seems to wander at random within a bounded region of phase space! With a further increase in E_0 the region occupied by random motion widens and the one occupied by simple motion contracts, see Fig. 2.7b. Thus we realize that already the motion of two coupled nonlinear oscillators within a rather simple model may be very complex.

---

5 The problem is that (*1*) a result due to an approximate method is not yet a (physicist's) proof while (*2*) the averaging methods enforce a specific answer.

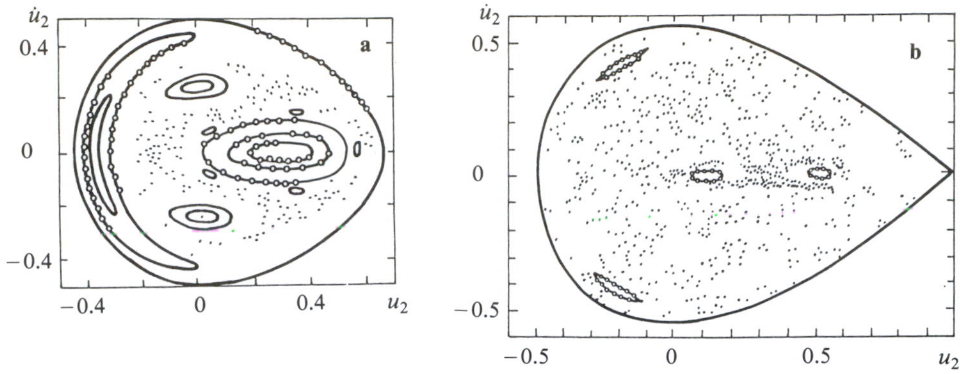

Fig. 2.7a,b. The complicated motion of the system (2.8) of two nonlinear oscillators with
a) $E_0 = 0.125$ and
b) $E_0 = 0.167$, see [2.9]

Where does this complexity come from? We shall attempt to answer this question later on in Chap. 3, *Chaos*.

Let us proceed on our way towards the study of complicated dynamic objects and consider now not just three coupled nonlinear oscillators, but a lot of them, a complete ensemble. First we shall consider two limiting cases: The one in which the coupling between the oscillators is very strong and ordered (for instance, nonlinear crystals and lattice models of continuum media) and the alternative in which the coupling between the oscillators occurs only due to an external field [into this class of problems fall questions related to collective effects in electron beams, electromagnetic waves and many others related to the interaction of fields with matter). Let us now start with the latter.

2.1.3 Nonlinear Landau Damping and Amplification

A very old problem relates to the behaviour of a large number of oscillators, e.g. oscillators in the field of a periodic wave. Already before Maxwell, there appeared a theory of the dispersion of light waves relying on a model with oscillators embedded in an elastic ether. Then the classical electron theory emerged, the theory of sound wave dispersion by gases and of electromagnetic waves in the ionosphere, etc. All of these are problems of the behaviour of an ensemble of linear oscillators. Where would nonlinearity lead them to?

Let us consider the effect of the nonlinearity of an individual oscillator using the example of the interaction of a travelling wave with an electron flow. The corresponding velocity distribution $f(v)$ of the electrons is depicted in Fig. 2.8. In a coordinate system moving with the sinusoidal wave $E(x,t) = \varphi_o \cos(\omega t - kx)$ all particles can be classified as being trapped or being in transit. The particles with speeds in the range $\omega/k \pm \sqrt{e\varphi_0/m}$ do not have enough energy to overcome the potential barrier $e\varphi_0$ so that they oscillate in the well of the wave. The particles with speeds outside this interval take almost no notice of the wave, see Fig. 2.9. Each i-th electron in the field of the sinusoidal wave behaves like a nonlinear oscillator or a pendulum

$$\ddot{x}_i + \omega_0^2 \sin x_i = 0 \ , \quad i = 1, 2, \ldots, N \ , \quad \omega_0^2 = k^2 e\varphi_0/m \ . \tag{2.10}$$

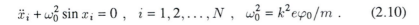

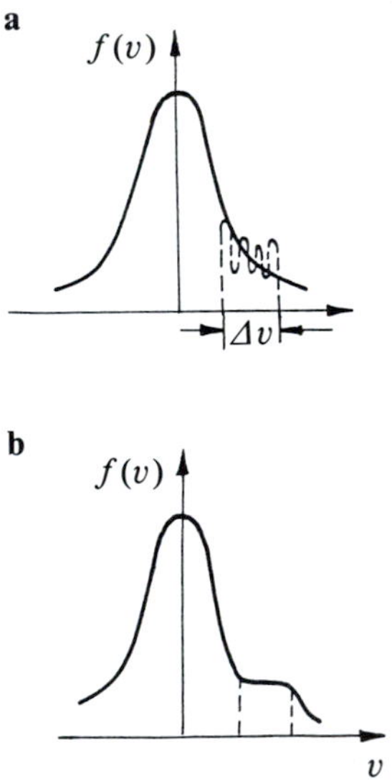

Fig. 2.8a,b. The velocity distribution function of electrons:
a) oscillations in the field of a periodic longitudinal wave;
b) formation of a plateau

Oscillations of the pendulum correspond to trapped electrons and rotations to transiting electrons, see Fig. 2.9. Thus the particles in the field of the wave constitute an ensemble of identical nonlinear oscillators that differ only in their initial energies.[6] How will such an ensemble behave in time? This depends on the energy or velocity distribution function $f(v)$ of the electrons. Since the interactions between the oscillators and their influence on the wave are not yet taken into account, this question is answered quite simply by considering the motion of the electrons in the phase plane. Figure 2.9 gives the dependence of the potential ϕ in which the particles move as a function of the coordinate x and depicts the phase trajectory of the electrons during their motion in the wave. The electrons with small speeds are trapped by the wave,[7] while the ones with sufficiently large speeds transit over it. In the $(\dot{x} = v, x)$-plane the separatrix, denoted in Fig. 2.9 by a dashed curve, separates trapped and transit electrons. The separatrix itself corresponds to the motion of electrons transiting in an infinite time from one potential hump to another one.

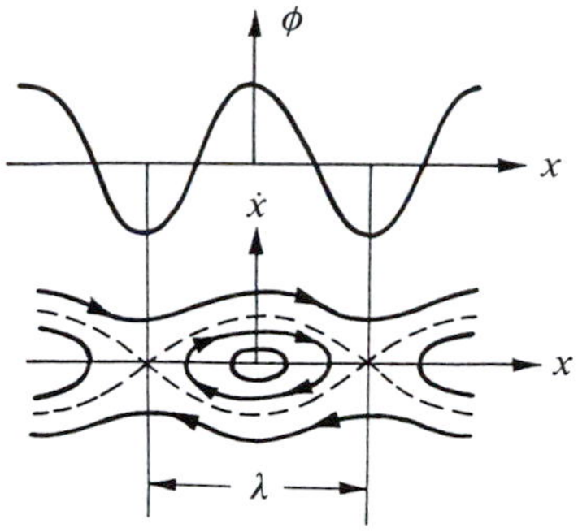

Fig. 2.9. Phase portrait of a nonlinear oscillator describing particles being captured and passing through a periodic field

Let us now consider the time variations of the velocity distribution function $f(v)$ for resonance particles (that are located in the region of the (v, x)-plane bounded by the separatrix and in its vicinity). We shall assume that this is for $t = 0$ an ordinary Maxwell distribution, see Fig. 2.8a. The motion of the electrons in the wave field corresponds to the one of a frictionless pendulum, implying that we deal here with conservative oscillators. For the flow of the phase fluid in the (v, x)-plane describing the evolution of $f(v)$ this implies that it should be incompressible whence it follows that the value of f is constant along the phase trajectories. We should bear in mind that at $t = 0$ amongst the trapped particles there are more slow electrons with speed $v < \omega/k$ than fast ones; we have $\partial f/\partial v|_{v=\omega/k} < 0$. At $t = 0$ the greater portion of the trapped particles will be in the lower halves of the "cat's eyes" in the (v, x)-plane, see the hatched area in Fig. 2.10a. Due to the nonisochronism of motions along the trajectories (the electrons move very slowly near the separatrix and rotate quickly in the vicinity of the point $(0, 0)$) the trapped particles in the central part will rotate faster. After half a period the picture will be as shown in Fig. 2.10b. Thus the upper part of the "cat's eye" will contain more particles than the lower one implying that the derivative $\partial f/\partial v$ for the trapped particles will change its sign. Because of the rotation such a change of sign will take place every half-period, see Fig. 2.10c. As time evolves the initial region will take on the form of a twisted spiral with the number of its turns increasing continuously. If we wait long enough, all oscillators will reassemble in the initial phase volume since the motion of a conservative system (2.10) of an arbitrarily large number N of oscillators is reversible. In reality, however, it is obvious that

[6] For simplicity we assume that the oscillators are on the average homogeneously distributed in space.

[7] Since we went over to the moving coordinate system of the running wave, our considerations include only those electrons that have speeds close to the ones of the wave. In electronics they are referred to as resonance electrons.

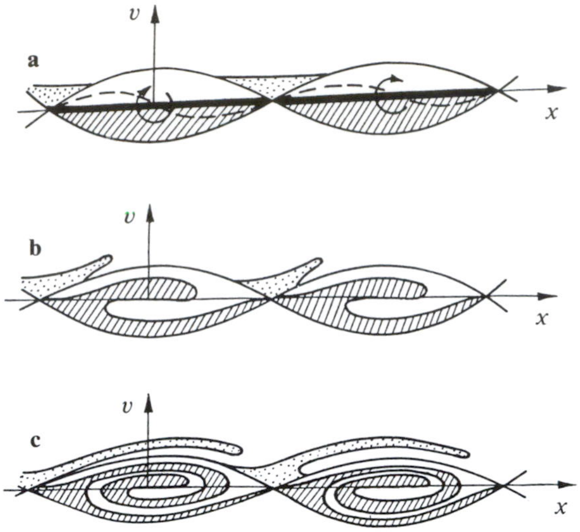

Fig. 2.10a–c. Evolution of the phase volume in an ensemble of noninteracting electrons (oscillators): *a*) $t = t_1$; *b*) $t_2 > t_1$; *c*) $t_3 > t_2$

no miracle will occur – no matter how long we wait: no matter how weak their (Coulomb) interaction is, it will give rise to particle mixing so that they will homogeneously fill the entire region within the separatrix or "cat's eye". In this process the number of trapped particles moving faster than the waves ($v > 0$) will on the average become equal to the number of particles moving slower ($v < 0$) so that the distribution function will acquire the plateau indicated in Fig. 2.8b. Since the average kinetic energy of the captured particles increases in such a mixing, the sinusoidal wave in which the particles are oscillating has to lose some energy for "heating" them up. This energy loss of a monochromatic wave is often referred to as *nonlinear Landau damping* [2.10].

Actually, this phenomenon cannot be understood and described within quasi-linear physics; only strongly nonisochronous oscillators (defined by the presence of separatrices) give rise to mixing.

Nonlinearly oscillating electrons in a harmonic potential are an example of an ensemble of nonlinear oscillators. Let us now consider another limiting case of a system consisting of a large number of nonlinear oscillators, namely ordered chains.

Examples of such chains are displayed in Fig. 2.11. If the individual oscillators in the chain were linear ones, then passing over to normal coordinates would again lead to an ensemble of uncoupled, but *non*identical "normal" oscillators representing collective oscillations of the chain. Approximately the same can be done for a strongly nonlinear system, however, in this case the new "collective" oscillators (or modes) will already weakly interact with each other. It might appear as if we reduced the problem to the preceding one, namely the analysis of interacting quasi-harmonic oscillators. This interaction is effective if the oscillator frequencies obey resonance conditions like (2.6). Possible consequences are either a periodic energy exchange between the oscillators or equilibrium between them (for specially chosen oscillation amplitudes and phases). However, even for weak nonlinearities

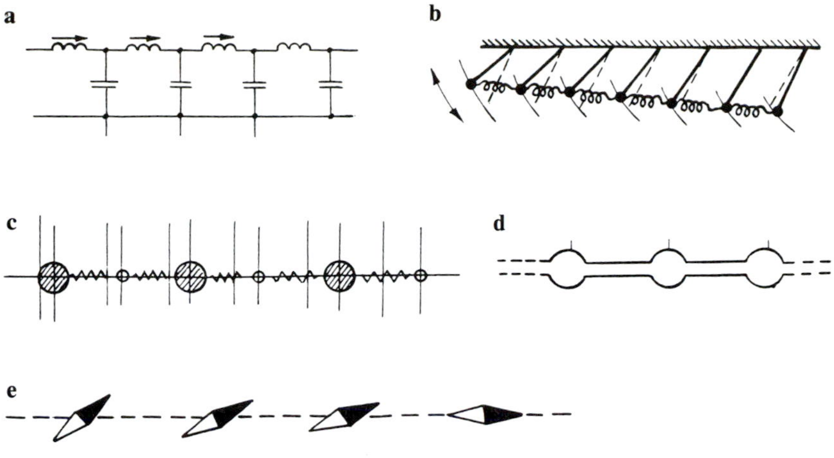

Fig. 2.11a–e. Chains of coupled nonlinear oscillators:
a) nonlinear LC-chain;
b) chain of pendulums coupled by springs;
c) one-dimensional "lattice" of particles with different masses;
d) acoustic analogue of a one-dimensional chain;
e) a chain of compass needles

our quasi-linear experience fails usually if the number of oscillators N is very large $N \gg 1$.

Let us have a look at a particular example:
We consider acoustic oscillations in a tube filled with air and interconnect beginning and end of it to obtain thus an acoustic ring resonator. Under certain assumptions such a resonator has an equidistant spatial excitation spectrum with $k_n = 2\pi n/L$ for $n = 1, 2, \ldots$ where L is the length of the ring. In view of the absence of dispersion for acoustic waves, the spectrum of the eigenfrequencies (normal oscillations) of the resonator will also be equidistant. They will be multiples of each other $\omega_n = n\omega_1$. We shall now take into account an arbitrarily weak nonlinearity of the medium which to a first approximation will be quadratic in the field. If the total initial energy were concentrated in the first mode with ω_1 then it would gradually be pumped into the mode with $\omega_2 = 2\omega_1$, see (2.5). Subsequently, the combined action of the modes with ω_1 and ω_2 will excite the third mode with $\omega_3 = \omega_1 + 2\omega_1$, etc. An intuitive guess about the future of the system would be extremely difficult. That holds even for the excitation of a larger and larger number of oscillators or modes with very weak nonlinearity. "The effect of a small nonlinearity of a large number of oscillators gives rise to indefiniteness." To resolve this indefiniteness without an a priori hypothesis it is inevitable to solve the initial nonlinear problem outside the realm of quasi-linear physics.[8] Such an attempt was undertaken by Fermi, Pasta and Ulam in 1952. The basic ideas will be recalled in the following section.

[8] Let us give way to speculations and list some possible answers: (*1*) after a sufficiently long time the nonlinear frequency shift will give rise to a phase shift between the oscillators, they will get out of resonance with each other leading again to an ensemble of uncoupled weakly nonlinear oscillators with independent frequencies; (2) the phases of all oscillators will be synchronized and something like a δ-pulse will emerge; (*3*) all modes or oscillators will be split up into groups (within the respective groups they will be synchronized) and different groups will interact with one another like strongly nonlinear systems. Other possible speculations are left to your imagination.

2.2 Solitons

In this section we shall deal with the particle-like solitons, solitary waves and their "relatives" the shock waves.

2.2.1 The Fermi – Pasta – Ulam Paradox

In connection with the problems of heat conductivity and heat capacity of crystalline solids the behaviour of chains of coupled oscillators was already studied at the beginning of this century. It is well-known that each normal mode (of oscillation) corresponds within classical physics to the energy kT and within quantum mechanics to $h\nu/(\exp[h\nu/kT]-1)$. But why is such a universal energy distribution per degree of freedom established for arbitrary initial conditions? How does thermalization occur? – These questions were of interest to everybody dealing with the theory of heat capacity. It appeared natural to relate the possibility of thermalization to the nonlinearity of the oscillators used to model such systems. *Fermi, Pasta* and *Ulam* attempted to verify this point of view by direct numerical calculations [2.11] – nowadays we would say by "computer experiments". They investigated the behaviour of a chain of 64 nonlinearly coupled oscillators $i = 1, 2, \ldots, 64$ with

$$\ddot{u}_i + 2u_i = (u_{i+1} + u_{i-1}) + \alpha[(u_{i+1} - u_i)^n - (u_i - u_{i-1})^n] \ . \quad (2.11)$$

In the early 1950's at Los Alamos, Fermi, Pasta, and Ulam (FPU) studied numerically on a computer a one-dimensional lattice of many equal mass particles with weakly nonlinear nearest-neighbor interactions ...

Entirely unexpected, however, was the phenomenon now called "FPU recurrence". After many linear oscillation periods, the slowly evolving wave form eventually returned almost to that of the initial conditions! And this happened for various forms of the nonlinear term and for various initial conditions.

M. Kruskal
in "Nonlinear Evolution Equations"
F. Calogero (Ed.)
(Pitman, London 1978) pp. 3

In these calculations the index n was taken to be two or three. In these computer experiments they observed features *that surprised us from the very onset.* The system did not thermalize! Instead they observed at the start a transfer of the energy from the first strongly excited mode to higher modes and subsequently all the energy (with a precision of about 1% error) was re-deposited into the first mode. The chain exhibited a clean recurrence effect. But don't we observe the same periodic motion in a system with only two coupled nonlinear oscillators, see Fig. 2.3b? Besides, where is "the strong nonlinearity of the problem"? Well, the paradox itself does not at all appear to be paradoxial: since two or three resonantly coupled oscillators exhibit quasi-periodic energy exchange, why should it be a surprise to observe it also in a system with several tens oscillators of exactly the same type (or mode)?

Yet, there is something surprising here, and there is strong nonlinearity. We shall try to explain the situation. First we shall draw your attention to the fact that these chains are models of media with dispersion. The (spatial) dispersion is related to the discreteness of the chain and is strongest for high-frequency (short-wave) excitations. Owing to dispersion, the resonance conditions for frequencies are met less and less exactly as the number of oscillators (modes) increases – the resonant detuning δ grows. In equations for phases like (2.5) the detuning will appear on the right-hand side. Because of the detuning the phases of interacting modes (when they have grown in number, the energy of the fundamental mode will be transferred to many modes) will pulsate rapidly and the energy exchange between the modes will cease.[9] There is irreversibility! That is exactly what Fermi was looking for when he performed his computer experiment. A sceptic could argue: "All the same, the system is a Hamiltonian one. Consequently time reversibility enforces the initial state to be recovered." However, this would be in contradiction with our experience with real physical systems where a great number of interacting modes is understood to imply an extremely large recurrence time so that arbitrarily small dissipative effects must be an obstacle for recurrence. In numerical experiments such unaccounted for effects usually do manifest themselves so that the authors' surprise was quite natural when they failed to observe thermalization. It was still more surprising to find out that recurrence proved to be stable not only with regard to changes in the initial conditions but also with respect to an increase in the number of oscillators in the chain. (Later on computer experiments were performed for chains containing as many as 250 oscillators.) This gives rise to the thought that recurrence should as well be observed in the continuum limit (as $N \to \infty$), i.e. in the transition to a continuous system with dispersion.

Exchanging in (2.11) the variables $u_i(t)$ against their continuous counterparts $u(x,t)$ and using the Laurent series expansion, the *Boussinesq equation*

$$\partial_{xx}[\partial_{xx}u + \alpha u^2 + \gamma u] = \partial_{tt}u \ , \tag{2.12}$$

is seen to be the continuous analogue of (2.11). Originally (2.12) was derived for water waves. It has stationary wave solutions $u(x,t) = u(\xi = x \pm vt)$ running to the right or to the left.

Boussinesq derived (2.12) in 1872 in connection with shallow water waves. Its stationary wave solutions may travel to the right or to the left. Applications in fluid dynamics, plasma physics, etc.

The uni-directional analogue of (2.12) is the *Korteweg – de Vries equation* (KdV)

$$\partial_t u + u\partial_x u + \beta\partial_{xxx}u = 0 \tag{2.13}$$

describing gravity waves on shallow water. If Fermi, Pasta and Ulam had taken the nonlinearity of the chain to be cubic then they would have obtained the *modified* Korteweg – de Vries equation (mKdV)

Korteweg and de Vries derived their now famous equation (2.13) in 1895 for the (uni-directional) propagation of waves on the surface of a shallow canal. It now serves as the simplest possible nonlinear and dispersive equation for exploring phenomena ranging from nuclear physics to meteorology and astrophysics.

$$\partial_t u + u^2\partial_x u + \beta\partial_{xxx}u = 0 \tag{2.14}$$

as the respective uni-directional continuous analogue of (2.11). Stationary waves in systems described by (2.12) or (2.13, 14) are known to be represented by trajectories in the phase plane $(du/d\xi,\ u)$, see Fig. 1.1. The particular trajectory called separatrix going from one saddle to another one corresponds to a solitary wave or soliton, see Fig. 1.1b. Solitons are strongly nonlinear solutions which can not be found within a quasi-linear approximation. Solitons are intimately related to the Fermi-Pasta-Ulam paradox.

The modified Korteweg – de Vries equation (2.14) is discussed in electric circuit theory, double layer theory, multi-component plasmas.

[9] These or similar arguments are usually employed in the derivation of kinetic equations for the mode intensities $N\omega_i = |a_{\omega_i}|^2$ in the theory of weak turbulence [2.12].

Nonlinear LC chains (as depicted in Fig. 2.11) are to a good approximation described by (2.12) or (2.14) and may thus be used to explore the solutions of these equations. The results of related experiments are displayed in Fig. 2.12. Exciting the circuit sinusoidally one observes that the sinusoidal wave is transformed into a periodic sequence of solitons, i.e. a large number of oscillators or harmonics has been excited. This periodic structure of solitons lives fairly long, then the periodic train of solitons is again converted into sinusoidal waves – all harmonics return their energies to the first one. For the initial value problem almost the same behaviour is observed in the Fermi–Pasta–Ulam chain. If one excites at $t = 0$ (or $x = 0$) a soliton in the chain then it will propagate stably. Thus the Fermi–Pasta–Ulam chain may be compared with a nonlinear pendulum (moving in functional space) which passes during its motion through stable (solitons) and unstable (sinusoidal modes) equilibrium states. This is an intrinsic feature of fully integrable conservative systems. The existence of stable solitons is characteristic for such systems.

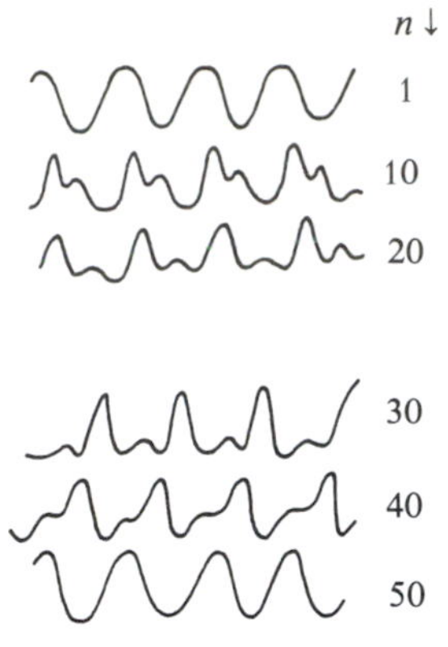

Fig. 2.12. Periodic evolution of nonlinear waves in LC-chains (with the number of cells increasing from the *top* to the *bottom*)

Put more rigorously, complete integrability of systems with N degrees of freedom implies that the systems have a complete set of N independent commuting integrals of motion. Almost all of the system's finite motions correspond to phase trajectories on N-dimensionial tori and to quasi-periodic time evolution.

This is indeed an irony of fate: even two coupled nonlinear oscillators behave at sufficiently high excitation energy (see Fig. 2.7) stochastically while we have here an entire chain of such oscillators that gives rise all of a sudden to an integrable or nearly integrable system! But integrable systems are just tiny "islands" in the space of all dynamic systems. It was an amazing combination of circumstances that lead Fermi, Pasta and Ulam to performing their numerical experiments on chains of the type (2.11). A slightly different coupling between the elements of the chain would have given rise to a "chaotic" system. For example, if the dispersion law of (2.14) is changed into a steeper function by replacing $\partial_{xxx}u$ by $\partial_{xxxxx}u$, stochasticity will be observed; even within the class of stationary wave solutions $u = u(\xi = x - vt)$. Such waves may be described by

$$\frac{d^4u}{d\xi^4} + \alpha\frac{d^2u}{d\xi^2} - vu + u^2 = 0 \ . \qquad (2.15)$$

As shown in [2.13] the phase space of this harmonic oscillator contains a region with complex behaviour, a homoclinic structure, Fig. 2.13. In a short historical note *Danilov* [2.14] elaborates on *Nonlinear Dynamics: Poincaré and Mandelstam.*[10] Figure 2.14a shows the numerical solutions of (2.15) and Fig. 2.14b the oscillograms observed in a nonlinear LC chain with the equivalent scheme shown in Fig. 2.15. Now we are in the position to answer the question as to why an increase in the number of oscillators in the chain (2.11) has practically no influence on its characteristics. The dispersion law of such a one-dimensional chain is given in Fig. 2.16. The number of particles determines only the density of the points forming this characteristic (i.e. the number

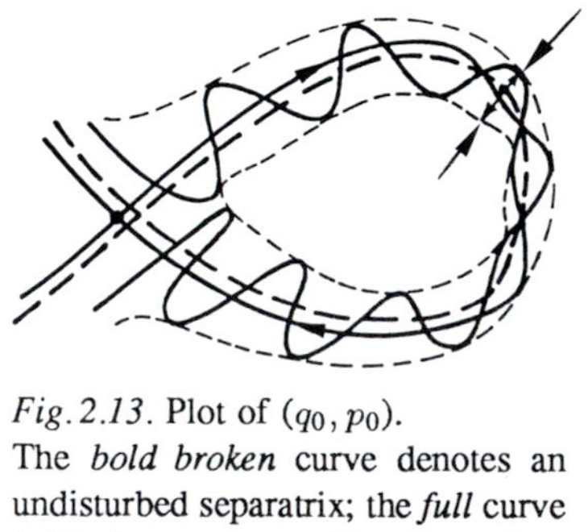

Fig. 2.13. Plot of (q_0, p_0). The *bold broken* curve denotes an undisturbed separatrix; the *full* curve (with the *arrows* on it) a homocline trajectory in its vicinity on a Poincaré section; the *thin broken* curve shows the boundary of the stochastic layer with the width d_{max}

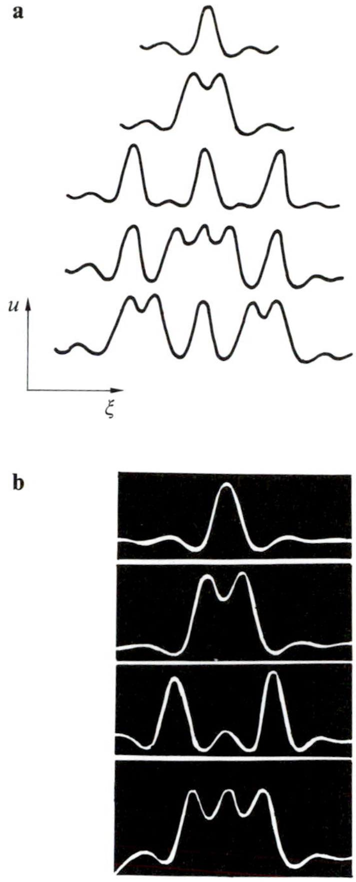

Fig. 2.14a,b. Multi-soliton solutions in a nonlinear chain:
a) numerical solutions;
b) real physical experiment

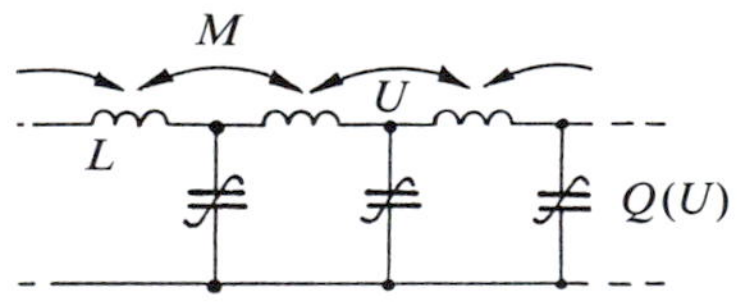

Fig. 2.15. Equivalent scheme of a nonlinear chain (see Fig. 2.14) supporting multi-solitons

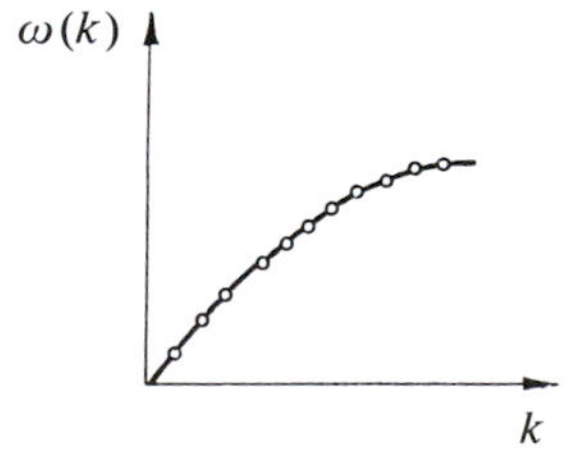

Fig. 2.16. Dispersion law for waves in the one-dimensional chain (2.11)

of normal oscillations of the chain) and has no effect on the shape of the curve – the type of dispersion remains unaffected. Excitation of a mode in a longer chain obviously increases only the recurrence time – the initial energy needs more time to get distributed amongst a larger number of normal oscillators, but the nature of the energy exchange between the modes is unchanged. It would not even change if the infinite quadratic chain were exchanged against the continuous medium described by the Boussinesq equation (2.12).

Unfortunately, we are not able to go into more detail here but despite the "low power" of the manifold of integrable systems, they play an exceptionally important part in the physics of nonlinear waves, both as specific examples from which certain general mechanisms of nonlinear phenomena can be inferred and as reference systems on the basis of whose solutions it is possible to construct approximate solutions of similar nonintegrable systems, see Chap. 3. Among such known nonlinear solutions solitons play a prominent role due to their amazing stability and the fact that they are actually *analytical* solutions of the underlying partial differential equations. Most localized solutions in integrable systems disintegrate for $t \to 0$ into a sequence of solitons, the "normal modes" of strongly nonlinear wave systems or media. As to be expected for normal modes they are stable and do not interact with each other in the absence of perturbations (including those that disturb integrability).

2.2.2 Solitons as Particles

The term "soliton" was coined by *Kruskal* and *Zabusky* in 1965 [2.19]. Investigating the phenomenon of recurrence in the Fermi-Pasta-Ulam problem they performed computer experiments with the KdV equation (2.13) with a very small value of the parameter β (weak dispersion).[11] They observed that the initially smooth wave becomes steeper and develops directly behind the wave front oscillations each of which grows with time t acquiring a stable amplitude and shape corresponding to the separatrix in the phase plane $(\partial_\xi u, u)$, see Fig. 1.1b. In other

[10] The homoclinic structure was discovered by *Poincaré* during his investigations of the three-body problem back in 1889. He wrote [2.15]: *Their (separatrix) intersection gives rise to something like a lattice or fabric like a net with infinitely close knots, – none of the curves will ever intersect with itself but it has to bend in a very complicated manner to intersect an infinite number of times with the meshes of the net. One has to admire the complexity of this figure which I do not even attempt to draw. Nothing else can give us a better representation of the sophistication of the three-body problem and all problems of dynamics which do not possess homomorphic integrals – implying that the Bellin series diverges.* A complete description of the trajectory within this structure has been given rather recently [2.16–18]. In particular, it was revealed that such a structure contains an even set of unstable (saddle) periodic trajectories, between which the "oscillator" wanders (for a wide range of initial conditions), see Fig. 2.13.

[11] Note that for $\beta = 0$ the KdV equation transforms into the known ordinary wave equation. At $t = 0$ it possesses a continuous solution which – due to the action of the nonlinearity – "breaks" in a finite time to become discontinuous and multi-valued, see below.

words, it becomes a solitary wave since the steepening of the profile due to the nonlinearity is exactly compensated for by the spreading due to dispersion. Zabusky and Kruskal called these solitary waves "solitons" because they do not break or disperse when they interact not even when "passing through each other", see Fig. 2.17. In this word the ending "on", stresses the particle-like properties (electron, proton, nucleon, photon, etc.) of these peculiar spatially localized solutions of nonlinear fields. Basic model equations in soliton theory are nonlinear evolution equations like the Korteweg-de Vries equation (2.13), the *Klein-Gordon equation* (KG)

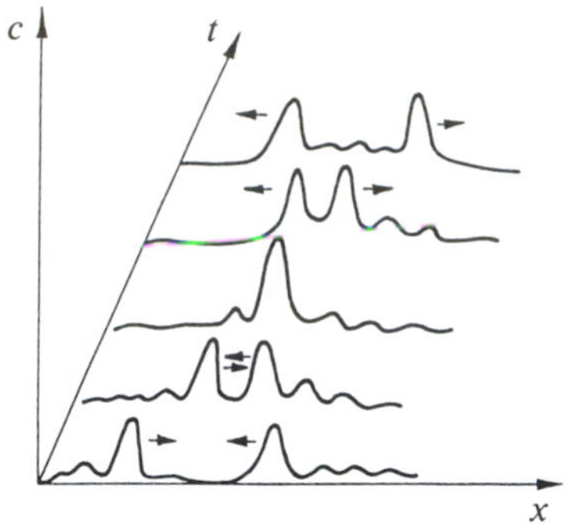

Fig. 2.17. Scattering of ion-acoustic solitons (the density is denoted by c)

$$\partial_{tt}u - \partial_{xx}u + f(u) = 0 \tag{2.16}$$

(with applications in field theory), the *nonlinear Schrödinger equation* (NLSE or NOSE)

$$i\partial_t u + \Delta u + \kappa|u|^2 u = 0 \tag{2.17}$$

(with applications in various branches of science including quantum mechanics) and others [2.19–21]. Taking in these equations high-frequency dissipation into account they are seen to support solutions in the form of shock waves, see below.

Knowing the solitons – elementary localized excitations of the nonlinear system – one can proceed to explore still further a wide range of phenomena; as usual one may again employ perturbation theory but at a different level. Thus, it is possible to describe the weak interaction of solitons with each other and with non-solitonic excitations [2.22, 23], their propagation in nonuniform media (e.g. in the study of tsunamis [2.24]), the behaviour in external fields and so on. For weakly interacting solitons (or cnoidal waves) the – at first sight rather superficial – analogy with particles turns out to be very profound. If through the entire process the velocity or energy difference (which is in this case the same) between the solitons is small and the distance between their maxima remains large compared to their effective width throughout the entire process then their interaction is literally analogous to the one of particles obeying Newton's equations. A soliton in the field of the tail of another soliton behaves like a marble in a chute. For example, for a pair of solitons we obtain [2.22, 23]

$$\frac{d^2u}{dt^2} - v'_E f(v, u) = 0 \tag{2.18}$$

where u is the distance between the soliton maxima, $f(v, u)$ is the force field of the "tail" of one soliton at the position of the other and $v(E)$ is the soliton speed as a function of energy. For small interactions, equations of the form (2.18) can be derived from the initial wave equations by representing the field in the neighbourhood of each soliton (whose parameters are assumed to change slowly) in terms of an asymptotic series and by requiring boundedness of the terms of this series.

Once the soliton-particle analogy has been established (i.e. once (2.18) has been obtained), all that is needed to describe the soliton interaction is the form of the force function $f(u)$, that is the nature of the soliton tails. If $f(u)$ is a monotonic function, the solitons are either repelled or attracted; naturally, (2.18) is no longer valid when their fields overlap strongly. However, if the solitons have oscillating tails, as in the case of capillary-gravity waves in shallow water [2.25], or in a nonlinear transmission line with inductive coupling between its elements [2.23], then $f(u)$ changes its sign and the solitons are alternatively repelled and attracted to form an oscillating pair (a bound state, see Fig. 2.18).

The interactions of a large number of solitons of the same type can be analysed in a similar fashion, because the nature of the tails does not depend on the number of solitons sitting on them.

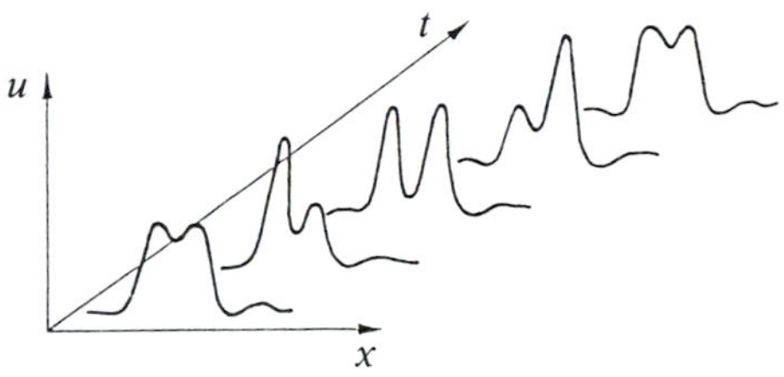

Fig. 2.18. Oscillating pair of (bound) solitons

2.2.3 Solitons and Shock Waves

In the presence of energy loss in chains like (2.11) or in their continuous analogues (2.13) or (2.14) it is strictly speaking impossible to have solitons the image of which in phase space would be given by the separatrixes (describing stationary waves). However, one should find their "relatives". Soliton-like structures in dissipative nonlinear media were actually known in physics long before the current strong interest in solitons. Reports on the investigation of quasi-stationary shock waves in gases appeared already back in the thirties [2.26]. In the fifties shock waves were studied in plasma physics [2.27] and magnetic hydrodynamics [2.28, 29]; research was carried out on electromagnetic shock waves [2.30] which are not significantly influenced by microscopic motions in the medium and variations in the thermodynamic state; more recent contributions discuss electromagnetic shock waves in vacuum [2.31].[12]

Shock waves are regions of fast changes or jumps of the parameters of the nonlinear medium or field which result from the evolution of a smooth perturbation. A well-known example is the (density) compression in an explosion. At first the formation of such jumps in homogeneous media with "good", i.e. sufficiently smooth, inital conditions was completely unexpected so that the phenomenon was referred to as the "gradient catastrophe". But still in the forties the nature of the phenomenon was unravelled and related to the absence or rather smallness

[12] The corresponding states of the electromagnetic field in vacuum are no longer described by the Maxwell equations that are linear in vacuum, but by different models of nonlinear field theory that take into account the interaction between the highly dense photons.

of dispersion in the nonlinear medium giving rise to discontinuities or jumps in the variables. – For weak dispersion all harmonics forming the initial perturbation move practically with the same speed; due to the generation of new harmonics the energy is transferred to higher modes in the spectrum thus giving rise to infinitely small wavelengths corresponding to jumps in the field parameters. Using the spatio-temporal language we could say that the wave profile becomes locally increasingly steeper until, finally, a discontinuity is formed. Even very weak "high-frequency" dissipation and dispersion are significant in the region of the discontinuity. Together with nonlinearity they lead to the onset of stationary shock waves.[13]

The evolution at the front of a stationary one-dimensional shock wave propagating with the velocity $v = v_p$ depends only on one coordinate $\xi = x - v_p t$. Such waves are described by ordinary nonlinear differential equations and can therefore be investigated in detail. In phase space the front of the shock wave corresponds to a phase trajectory (separatrix) connecting two equilibriium states (the state of the medium or field before the jump or discontinuity with the one after the jump, see Fig. 1.1a). For a unique trajectory given initial and boundary conditions yield a unique solution for the jump whose structure or wave profile is determined by this very trajectory.

In conservative systems analytical solutions of similar fame are: in linear media the harmonic oscillator, in nonlinear media the soliton (and in the presence of dissipation its sister, the shock wave).

Some initial data, for example in magnetic hydrodynamics, may correspond to several discontinuous or jump solutions. However, only those of them that depend continuously on initial and boundary conditions (the evolution requirement) are realized. Nonevolutionary discontinuities (a small perturbation of the initial data causes finite changes in the solution) are unstable with regard to disintegration. It is natural that the jumps formed as the result of a continuous steepening of the profile of a simple wave are always evolutionary ones.

The evolution conditions coincide with the uniqueness conditions for the structure of a stationary shock wave [2.32]. This remarkable circumstance explains why shock waves are stable, independent of the details of their front structure. They are determined by the dispersive and dissipative characteristics of the medium in the region of fast changing variables. Thus, if at the front the energy loss plays a decisive role, the wave profile will be smooth, Fig. 2.19a. If dispersion is important and dissipation negligible, then the front will have a very steep profile. For small dissipation the wave will consist of a train of nearly equal and almost independent pulses, see Fig. 2.19b.

In the limit of vanishingly small dissipation, the distance between the pulses at the front of the shock wave grows without limits (so that the term "wave front" loses its meaning) while the energy transfer to the pulse tends to zero. Such solitary pulses are solitons. In the phase space of stationary waves their image is the separatrix originating from the equilibrium state of the "saddle" type and returning to it, see Fig. 1.1b.

[13] In order to obtain such waves experimentally appropriate boundary conditions have to be created.

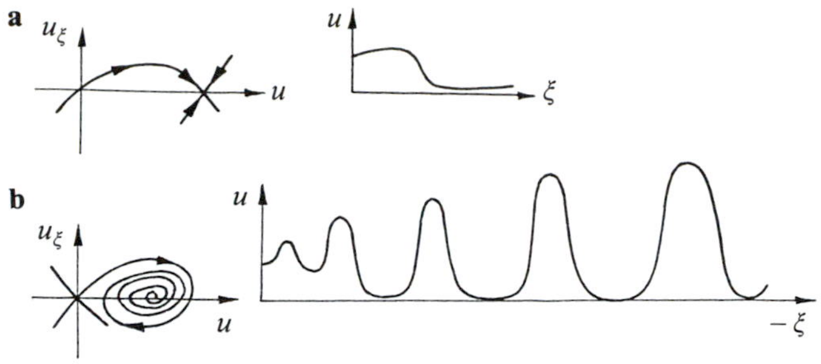

Fig. 2.19a,b. Stationary shock waves: *a*) weak dispersion in front of the wave; *b*) structure of the front of a stationary wave in a medium with dispersion and vanishingly small dissipation

In weakly dissipative media the solitons arising due to external energy sources are examples for autowaves and self-structures (see Chap. 4, *Structures*). Such objects can be described by the aid of a new nonlinear pertubation theory based on the known nonlinear solutions of the unperturbed (fully integrable) system.[14] In recent years the possibilities for studying systems close to fully integrable ones broadened significantly. This has been due to remarkable achievements in mathematical physics leading to the inverse scattering transform based on the use of inverse problems [2.33, 34].

The energy supply to soliton-like localized structures in a dissipative medium explains, for example, the phenomenon known as the Great Red Spot of Jupiter. There are very convincing arguments in favour of the interpretation that the Great Red Spot is an anticyclonic vortex (a two-dimensional Rossby soliton [2.35, 36] rotating around its own axis in the direction opposed to the one of the revolution of the planet). Such vortices are modelled in laboratory. They are generated in a thin fluid layer on the surface of a rotating parabolic bowl. Owing to viscosity, the lifetime of anticyclones is finite. However, in the presence of shear flows such vortices occur due to developing shear instabilities (Helmholtz instability) and may be stationary beings, since viscosity losses are compensated for by the energy due to the average flow, see Fig. 2.20 and [2.37].

The properties of the medium like spatial dimensions, nonlinearity and dispersion law provide the conditions for the existence of the soliton as an elementary object. In particular, they determine its stability which specifies the adiabatic interaction of soliton and non-soliton part ("tails") or external fields. The necessary condition for the stability of solitons is essentially the following: one of the integrals of motion (e.g. the energy integral) must assume an extremum at the soliton solution. This is a consequence of the Lyapunov theorem [2.38]. Although the conditions sufficient for the stability of solitons have not yet been formulated, the estimation of the existence of the extremum of the integral appears to be useful for predicting instability.

[14] It is convenient to construct the unperturbed solution in the form of an ensemble of interacting solitons on the basis of stable solitons. However, their stability is not necessary, it suffices if the lifetimes of the solitons are larger than the characteristic time of the problem.

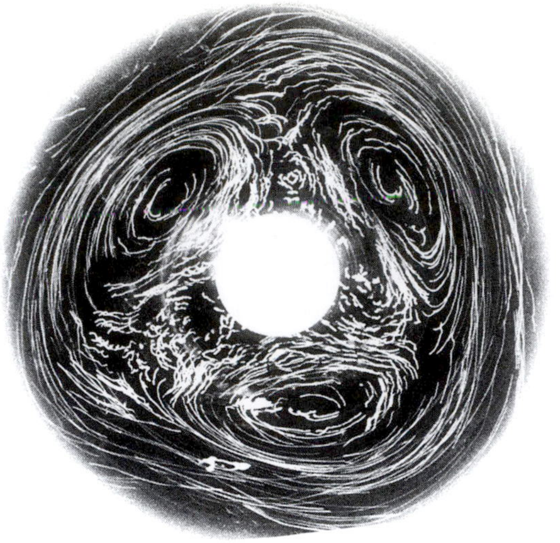

Fig. 2.20. Shear flow of a liquid in a bowl (with a parabolic profile of the bottom) which rotates around its vertical axis. The velocity difference corresponds to the formation of anticyclones. – The shown trajectories are taken by a camera rotating together with the bowl and made visible by white test particles floating on the surface of the water against the background of the black bottom; the white blot in the center is the disk drive used to generate vortices [2.37]

For example, in media with weak dispersion the propagation of multi-dimensional solitons is described by the *Kadomtsev-Petviashvili equation* (KP)

$$\partial_x(\partial_t u - 6u\partial_x u - \partial_{xxx} u) = 3\kappa\triangle_\perp u \ . \tag{2.19}$$

For $\kappa = 0$ it goes over into the Korteweg-de Vries equation which it generalizes by taking transverse diffusion into account. For negative dispersion $\kappa < 0$ (2.19) describes long gravity waves on (shallow) water, the propagation of sound in a solid, etc. For positive dispersion $\kappa > 0$ it models capillary-gravity waves, phonons in liquid helium and waves with a so-called decay spectrum (only waves having such a spectrum do form two- and three-dimensional solitons). The point is that in media with a nondecaying wave spectrum one-dimensional solitons are only stable with regard to transverse perturbations [2.39].

The soliton, i.e. the stationary solution of (2.19) of the form $u = u(x - vt, r_\perp)$ decaying into all directions, is determined by

$$[v\partial_{xx} + 3\kappa\triangle_\perp - \partial_{xxxx}]u = 3\partial_{xx}u^2, \tag{2.20}$$

with $v > 0$ (condition for the existence of a localized solution) [2.40]. The Hamiltonian form of (2.20) with $\kappa = 1$ is

$$H = \int \left[(\partial_x u)^2/2 + (\triangle_\perp \omega)^3 3/2 - u^3\right] dr \quad \text{with} \quad \partial_x\omega = u \ . \tag{2.21}$$

Using scaling transformations in two-dimensional space we can show [2.41] that the Hamiltonian acquires its minimum for a two-dimensional soliton. In other words, such a soliton is stable with regard to two-dimensional perturbations. In three-dimensional space there is no extremum, i.e. the soliton is unstable with regard to three-dimensional perturbations.

In order to understand the physics of soliton stability we must answer the following question for a medium without dissipation: how do perturbations distorting the soliton decay? There exist two typical mechanisms: the perturbations either go over from the soliton to the "tail" (radiation) or they are absorbed by the soliton thus increasing its energy somewhat but leaving it as a soliton. In each specific case under consideration the task is to determine which of these or similar mechanisms is the dominant one.

Stability and evolutionary characteristics of solitons as demonstrated, in particular, in the limiting transition from evolutionary shock waves, as well as their observation in very different experimental situations specified their "privileged" role in nonlinear physics. In particular, in the recent decades attempts have been undertaken to employ solitons for the description of strong interactions in quantum field theory [2.42]; it is the solitons that determine the propagation of dislocations in crystals as well as that of the magnetic flux across the continuous Josephson junction [2.43]. They appear as a result of wave self-focussing and self-modulation [2.44], correspond to wavepackets in deep water [2.45], and describe the behaviour of a non-ideal Bose-gas of weakly interacting particles [2.46]. Models emerged which relate solitons to collective excitations in molecular chains [2.47] etc. Much attention is presently paid to the investigation of the properties of ensembles of solitons (soliton "gas"). One hopes to be able to use these soliton models to prepare the grounds for the construction of a theory of strongly turbulent plasmas [2.41], towards the description of various "strong excitations" in solids and in other applications. There are contributions discussing not only the "kinetics" but also the "thermodynamics" of solitons [2.48].

We have started our discussion of solitons with the investigation of nonlinear chains of oscillating particles. Now it is – through the separation of the collective degrees of freedom (excitations) of such ordered ensembles of oscillators or their continuous analogues (nonlinear media) and through establishing the similarity of solitons with classical particles – that we are returning to finite-dimensional nonlinear dynamics. In a certain sense we may say that the circle has been completed.

... one of the most beautiful phenomena that I ever saw. The heap of water took a beautiful shape of its own; and instead of stopping, ran along the whole length of the channel to the other end, leaving the channel as quiet and as much at rest as it had been before.

J. Scott Russell
"Report on Waves"
Meeting of Brit. Soc. Adv. Sc.
(Murray, London 1844)

Solitons with oscillating tails will in interactions with each other oscillate (in space and/or time). If the dissipative energy losses are compensated for by external sources then we shall have self-excited oscillations. We will discuss them now.

2.3 Self-Excited Oscillations

Having made acquaintance with harmonic oscillations, it is anticipated that nonlinear oscillations with or without damping and amplification should exhibit more intriguing features than such idealized systems. A further step towards even more fascinating systems is the study of self-excited oscillations.

2.3.1 Examples and Definitions

What are the principally new phenomena to be discovered in dissipative nonlinear systems?[15] There are quite a few of them but the most prominent one is the generation of undamped oscillations with properties independent of the initial conditions, i.e. the generation of self-excited oscillations. Similarly to the case of solitons in conservative systems, we shall start the discussion of self-excited oscillations with a simple system with one degree of freedom, then we shall analyse two coupled self-excited oscillators, ensembles of self-excited oscillators and, eventually, non-equilibrium media. We shall distinguish between quasi-linear and strongly nonlinear problems that cannot be considered within quasi-linear physics.

Self-excited oscillations have always existed in nature and have been observed many centuries before man gave them this name. Let us recall, for example, the behaviour of a willow stalk growing at the bottom of a shallow river. "Excited" by a smooth current it oscillates with a definite frequency and amplitude! These oscillations are self-excited as a result of the interaction of the elastic stalk with the current. If there were no current, the stalk bent from its vertical position would move in a viscous fluid and return rapidly to the stable state of rest. The interaction with the liquid flow compensates the dissipative losses and the "stalk-current" system becomes a generator of oscillations.[16]

Thus, self-excited oscillations are undamped oscillations in a dissipative nonlinear system that are maintained by an external energy source and whose parameters (like amplitude and oscillation spectrum) do not depend on variations in the initial conditions but are intrinsic properties of the system. The term "self-excited oscillations" was introduced in 1928 by A.A. Andronov who related periodic self-excited oscillations to Poincaré limit cycles.

Self-excited oscillations differ in principle from other oscillatory processes in dissipative systems in so far that they are maintained without external oscillatory forces: the oscillations of a violin string when the bow is moved across it, of the current in a radio generator, of air in an organ pipe, of a pendulum in a clock – all of them are examples of self-excited oscillations. Self-excited oscillations arise as a result of collective instabilities with subsequent stabilization owing to suspended energy supply from a source or with progressing energy loss (dissipation). Figure 2.21a depicts a simple energy scheme for

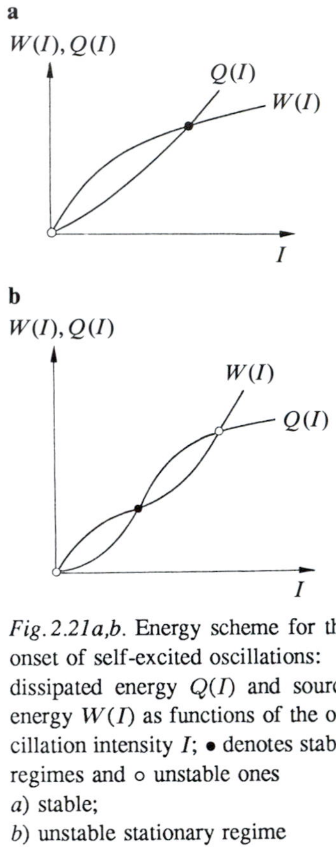

Fig. 2.21a,b. Energy scheme for the onset of self-excited oscillations: dissipated energy $Q(I)$ and source energy $W(I)$ as functions of the oscillation intensity I; • denotes stable regimes and ∘ unstable ones
a) stable;
b) unstable stationary regime

[15] A dynamical system is dissipative if (*1*) its phase space is unbounded and (2) there exists a bounded region in phase space into which the image point moves. For $t \to \infty$ the motion of a dissipative system corresponds in phase space to an attractor: the manifold of all image points belonging to it remains therein for all t from $-\infty$ to $+\infty$ and it attracts all trajectories located in its vicinity. Static, periodic and quasi-periodic regimes correspond to the simplest attractors: equilibrium states, periodic trajectories and tori, respectively.

[16] The excitation mechanism of the stalk surrounded by the flow of water is related to birth and separation of vortices from the stalk. A similar mechanism leads to the flapping of sails and flags.

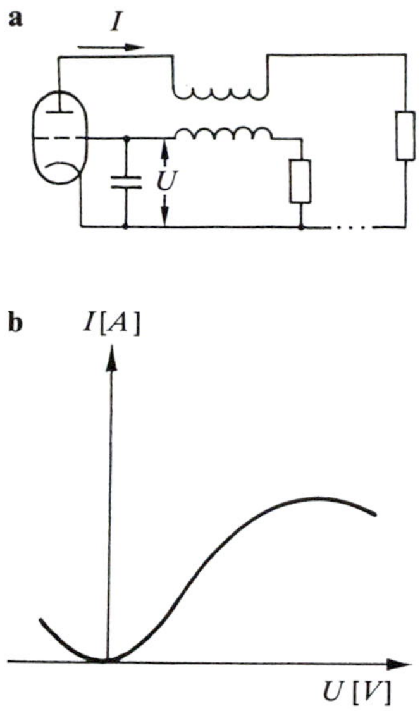

Fig. 2.22a,b. Electrical circuits: *a*) block scheme of a Van der Pol generator, $I = I_{\text{anode}}$, $U = U_{\text{grid}}$; *b*) characteristic of a triode; $I_{\text{anode}} = f(U_{\text{grid}})$

the onset of self-excited oscillations. The intensity I_0 of the regime of stationary self-excited oscillations is determined by the energy balance condition: the dissipative energy loss $Q(I)$ is exactly compensated by the energy supply $W(I)$ from the source, $Q(I_0) = W(I_0)$. If for varying I the energy loss $Q(I)$ changes faster than the energy supply $W(I)$ in the neighbourhood of the stationary regime, then this regime of self-excited oscillations is stable from the energetic point of view. If $W(I)$ changes faster, the stationary regime is unstable, see Fig. 2.21b.

The functions Q and W generally depend not only on the intensity of the oscillations but also on their phases so that the energy method cannot be used for reliable estimates of the stability of self-excited oscillations. Systems in which oscillations appear "spontaneously" without an initial external impulse are called systems with soft excitation. Systems that need a finite initial impulse to initiate self-excitations are referred to as systems with hard excitation. In the simplest self-excited oscillating systems one usually finds an oscillating system with damping, an amplifier, a nonlinear limiter and a feedback unit. For example, in the Van der Pol generator (Fig. 2.22a), the LC-circuit containing a capacitor C, an inductance L and a resistor R is a dissipative oscillator circuit (cathode-grid circuit and the inductance L' form a feedback unit). Self-excited oscillations in such a generator are established as follows: random disturbances excited in the LC-circuit through the coil L control the anode current of the tube which is an amplifier. With a positive feedback (given the appropriate location and choice of the coils L and L_1) energy is pumped into the circuit. If this energy is larger than the energy loss in the circuit, the amplitude of initially small oscillations in the circuit will grow. The anode current depends nonlinearly on the grid voltage, see Fig. 2.22b. Therefore we observe for growing oscillation amplitudes a decrease in the energy entering the circuit and at a certain oscillation amplitude the energy injected into the circuit will become equal to the energy loss. As a result, a regime of steady-state self-excited oscillations is established in which the external source (anode battery) compensates the energy loss.

These qualitative considerations lead us to the conclusion that self-excited oscillating systems must be essentially nonlinear. The nonlinearity governs the energy input from the source and the energy loss and limits the growth of oscillations.

To answer the question with regard to the nature of self-excited oscillations and to determine the dependence of their amplitude and shape on the parameters of the system, we shall analyse an appropriate mathematical model. Such a model for the simplest nonstatical attractor (Fig. 2.23) is the Van der Pol equation[17]

[17] This equation is written in the dimensionless variables: $x = \sqrt{\alpha}u$, $t = \omega_0 t_{\text{old}}$, $\mu = \gamma\omega_0$, where $\omega_0 = 1/\sqrt{LC}$ is the eigenfrequency of the oscillator circuit; $\gamma = \sqrt{LC}(MS_0 - RC)$ describes the excess of the energy supply over the generation threshold (for $\alpha > 0$ the losses in the circuit are larger than the injected energy), and $\alpha = 2MS_2/(MS_0 - RC)$ describes the amplitude of the self-excited oscillations.

$$\frac{d^2x}{dt^2} - \mu\left(1 - x^2\right)\frac{dx}{dt} + x = 0 \tag{2.22}$$

which is obtained by neglecting the grid current of the tube and approximating its characteristic curve by the one shown in Fig. 2.22b. In this system self-excited oscillations are described by a second-order differential equation whose phase space is a plane which imposes certain restrictions on the shape of self-excited oscillations. In such a system only periodic self-excited oscillations are possible. Let us explain this. The geometric image of the self-excited oscillations established in the phase space of the system is an attractor, i.e. a trajectory (or a set of trajectories) located in a bounded region of the phase space and attracting all neighbouring trajectories. Since the trajectories cannot intersect on the phase plane, we have in such second-order systems only a most simple nontrivial attractor, a closed trajectory to which all trajectories in its vicinity tend. Such a trajectory is the Poincaré limit cycle. The limit cycle is the image (in phase space) of periodic self-excited oscillations. The shape of the cycle determines the shape of the self-excited oscillations or, using the spectral language, amplitude and phase of their Fourier harmonics.

For small μ with $\mu \ll 1$ both the losses in the circuit and the energy pumped into the system are very small, (2.22) being close to the equation of a harmonic oscillator. The oscillations will be nearly sinusoidal with the frequency ω_0. The condition of energy balance merely "selects" among the set of closed trajectories the one on which the dissipative energy loss is compensated for by the energy supply and transforms it into a limit cycle without changing its shape, Fig. 2.23a. A detailed analysis of such quasi-harmonic self-excited oscillations can be performed employing, for example, the method of slowly varying amplitudes, where instead of the equations for x and dx/dt, the equations for amplitude A and phase ϕ of the harmonic oscillations $A(t)\cos[\omega_0 t + \phi(t)]$ are constructed and analysed, namely

$$\dot{A} = \mu A\left(1 - \alpha A^2 3/4\right) , \quad \dot{\phi} = \mu\beta A^2 \tag{2.23}$$

The symbol β denotes the reactive nonlinearity related to the nonlinearity of the ferrite in the coil L or the nonlinearity of the p-n–junction of the capacitor C. The stable equilibrium state of the averaged equations $A_0^2 = 4/3\alpha$; and $\phi_0 = \text{const}$ corresponds to a stable limit cycle. *Mandelstam and Papaleksi* [2.49] showed that such an approximate solution of (2.22)

$$x(t) = A_0 \cos(\omega_0 t + \phi_0)$$

is similar to the unknown exact solution not only in the finite time interval $T \sim 1/\mu$, but also in an infinite interval, i.e. as $t \to \infty$.[18]

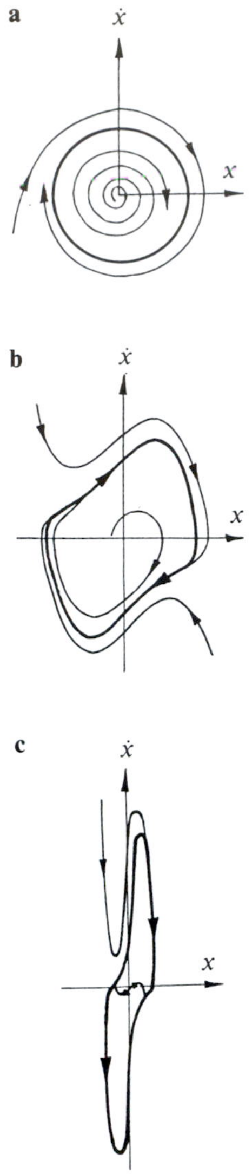

Fig. 2.23a–c. Phase portraits of Van der Pol generators with different values of the nonlinearity:
a) $\mu = 0.1$: quasi-harmonic oscillations;
b) $\mu = 1$: strongly nonsinusoidal oscillations;
c) $\mu = 10$: relaxating oscillations

[18] It should be noted that the nontrivial physical approach to the proof of a purely mathematical problem that was used in this paper was very fruitful, in particular, in justifying similar approximate methods in the theory of nonlinear waves [2.50].

In the other limiting case ($\mu \gg 1$), the loss in the circuit and the externally supplied energy will be very large compared to the stored energy, therefore the oscillations will be strongly non-sinusoidal, see Fig. 2.23c. Changing the variables in (2.22) we can now write it in the form of an equation with a small parameter ($\sim 1/\mu^2 = \varepsilon$) as the coefficient of a higher derivative. In the analysis of such self-excited oscillations it is convenient to split the motion up into a fast and a slow regime (with $\varepsilon \neq 0$ and $\varepsilon = 0$, respectively).[19]

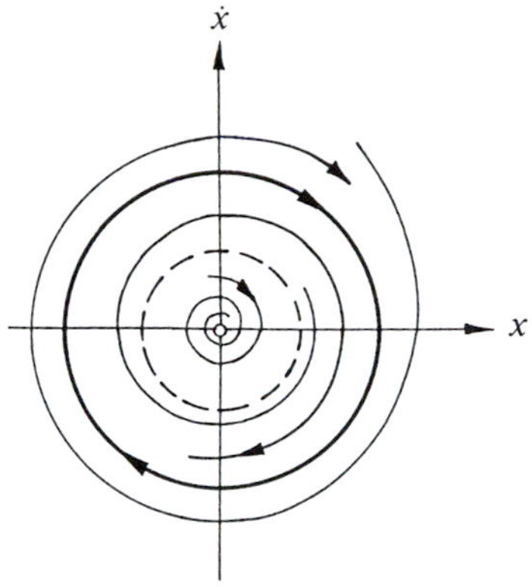

Fig. 2.24. Phase portrait of a generator with hard excitations

We emphasize that variations in μ do not lead to qualitative changes in the structure of the breaking up of the phase plane of (2.21) into trajectories. In the system there exists for any $\mu > 0$ a single equilibrium state ($x = 0,\ dx/dt = 0$) that is unstable and a single limit cycle that is stable. Qualitative changes (bifurcations) occur only when μ changes its sign, see below. This picture corresponds to the soft excitation of self-excited oscillations. The phase portrait shown in Fig. 2.24 depicts hard excitation: the oscillations in the system increase when there is an initial external impulse – the image point is outside the region of attraction of the stable equilibrium state (since the unstable limit cycle marks the boundary of the basin of attraction).

Practically all the experience of classical theory (at least for systems with strong nonlinearity) was related to the analysis of self-excited oscillations in the phase plane. That is why one associated the possibility to establish periodic motions corresponding to limit cycles exclusively with dissipative systems where undamped oscillations occur only at the expense of nonperiodic energy sources. Just a few years ago no-one would have thought of applying the term "self-excited oscillator" to a nonlinear oscillator with friction and a periodic (external) driving force:

$$\ddot{u} + \gamma\dot{u} - \alpha u\left(1 - u^2\right) = f_0 \sin(\omega t) . \tag{2.24}$$

However, it *is* a self-excited oscillator: such a nonlinear oscillator produces undamped oscillations whose parameters (intensity, frequency, and, in a more general case, its spectrum, etc.) do not depend on finite variations of the initial conditions and respond only weakly to changes in the external force.[20] In particular, in the nonautonomous phase space $\dot{u}, u, t$ of (2.24) there exist stable periodic motions to which (like to limit cycles of autonomous systems) there correspond stable stationary points (in the Poincaré map) if the system is observed stroboscopically at the period of the external force. We shall see in Chap. 3 that system (2.24) may also exhibit quite unusual properties: it can act as a self-excited noise generator.

[19] Similar ideas were later used in investigations of continuous nonlinear systems, specifically in deriving boundary conditions at a discontinuity in the theory of electromagnetic shock waves [2.30].

[20] This is the main feature of physical self-excited oscillators, rigorous definitions of them refer to (mathematical) models.

2.3.2 Competition and Sychronization

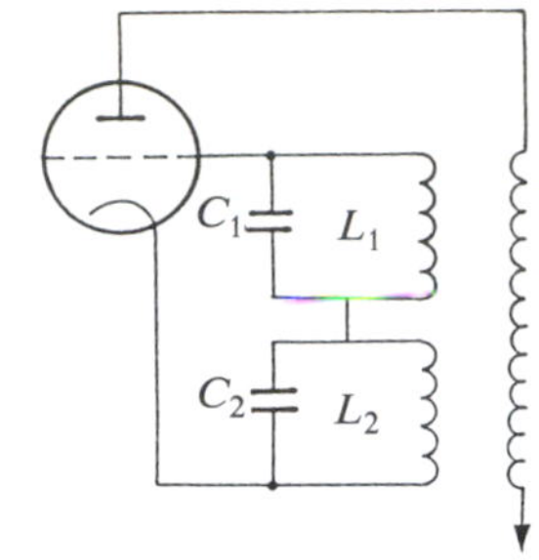

Fig. 2.25. Two-circuit self-generator

Figure 2.25 shows the scheme of the two-circuit vacuum tube oscillator that was investigated by Van der Pol and Andronov and Vitt almost half a century ago. They observed the most important effects characteristic of the interaction of two coupled self-excited oscillators like (2.22). The averaged equations for the complex amplitudes of coupled quasi-harmonic self-excited oscillations with incommensurate frequencies (for the scheme shown in Fig. 2.25 the ω_j with $j = 1, 2$ are the normal frequencies of the linearized system) have the form

$$\dot{a}_j = \mu h_j \left[1 - \alpha_j \left(|a_j|^2 + \sum_i \varrho_{ji} |a_j|^2 \right) \right] a_j \ . \tag{2.25}$$

In the general case with N normal frequencies we have $j = 1, 2, \ldots, N$.

Figure 2.26 shows the phase portraits of this system for $N = 2$ and various values of the parameters. The curves illustrate the classical effects of mode competition, coexistence of oscillations and of hysteresis (dependence of the established motion on the initial conditions, see below).

Mode competition – the suppression of some modes by others – in self-excited oscillating systems is related to the fact that the competing modes draw the energy to compensate their dissipative losses from a common source. As a result, some modes give rise to additional nonlinear damping of others. Thanks to the effects of competition and mutual synchronization of oscillations (see below) self-excited oscillating systems with a large number of degrees of freedom (or even an infinitely large number of them for continuous systems) may convert initial noise into a regime of regular periodic self-excited oscillations. This occurs as fluctuations at different frequencies are increased by developing linear instabilities. The competition phenomenon is observed for strong mode coupling ($\varrho_{12}, \varrho_{21} > 1$). For equivalent modes and mutual coupling the generation regime of the mode that dominated initially is established, Fig. 2.26b. With a change in the initial conditions, the system goes over from one regime to another, with oscillations occuring only at a single mode. That is, the detuning has to be changed so that its values differ for the motion going "there" and coming "back" (hysteresis). The range of detunings in which the generation frequency depends on the prehistory is also referred to as the "pulling" range.

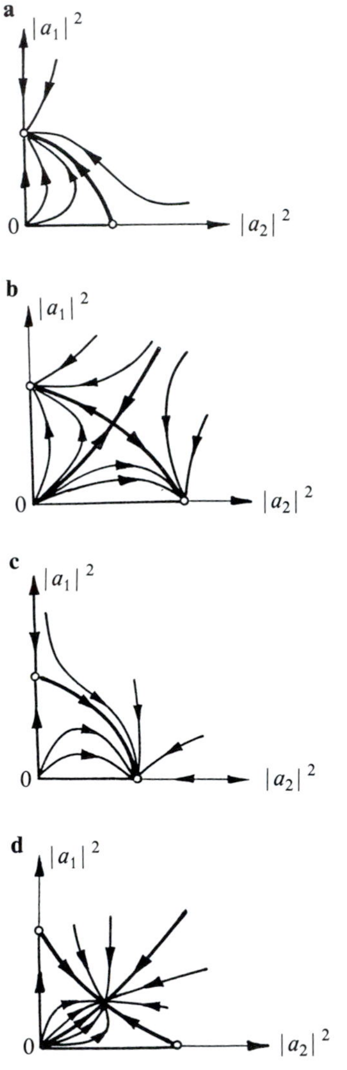

Fig. 2.26a–d. Phase portrait of the system (2.25) illustrating the effect of competing and coexisting oscillations ($N = 2$):
a) $\varrho_{12} > 1$ and $\varrho_{21} < 1$;
b) $\varrho_{12} > 1$ and $\varrho_{21} > 1$;
c) $\varrho_{12} < 1$ and $\varrho_{21} > 1$;
d) $\varrho_{12} < 1$ and $\varrho_{21} < 1$

In the interaction of quasi-harmonic oscillators with close or multiple frequencies synchronization is observed, i.e. mode-locking of the phases of interacting oscillations. We can distinguish between external sychronization of the frequency of the self-excited oscillator (the onset of oscillations with the frequency and phase corresponding to frequency and phase of the external periodic force) and mutual synchronization (the onset of periodic synchronized oscillations in an ensemble of subsystems with different frequencies in independent regimes). Frequency trapping is widely used for the control and stabilization of the frequency of powerful low-stability generators by means of high-stability low-power generators, e.g. in lasers. The trapping band (i.e. the region of detuning between the frequencies of natural oscillations and the external signal) inside which the synchronization regime is established, broadens with increasing amplitude of the external force. Outside the trapping band the stable regime of the excitation of periodic oscillations goes over into a regime of beats (quasi-periodic oscillations) or a stochastic regime, see Chap. 3, *Chaos*. The mutual synchronization of subsystems or different elementary oscillations (modes) is exploited when several generators are used to drive the same system, for the generation of short pulses in multimode oscillators, etc.

If the detuning between eigenfrequency ω_0 of the self-excited oscillations and frequency ω_{ext} of the external force is larger than the critical one $|\xi| = |\omega_0 - \omega_{\text{ext}}| > \xi^* = \xi_{\text{cr}}$ then the system is outside the synchronization region. Then the single-frequency regime is replaced by the double-frequency regime of oscillations. The system gives rise to beats. This process can be described by averaged equations for the amplitude and phase of quasi-harmonic oscillations of a nonautonomous generator. Representing the quasi-harmonic oscillations by $x(t) = A(t)\sin(\omega_{\text{ext}}t) + B(t)\cos(\omega_{\text{ext}}t)$ these equations are similar to (2.23). For slow amplitudes $A(t)$ and $B(t)$ they can be written in the form

$$\begin{aligned} \dot{A} &= \left\{A\left[1-(A^2+B^2)/4)\right] - B\xi + f_0\right\}/2\,, \\ \dot{B} &= \left\{B\left[1-(A^2+B^2)/4\right] + A\xi\right\}/2 \end{aligned} \qquad (2.26)$$

where f_0 is the amplitude of the external force. Analysing the onset of beats we are again concerned with a strongly nonlinear problem (in this case, however, at the level of an averaged description). The point is that system (2.26) has no small parameter, but for an understanding of the scenario replacing the trapping regime (with a stable equilibrium state in the phase plane (A, B)) by the regime of beats (with a limit cycle) we need a complete mapping of the trajectories on the (A, B)-plane. We must analyse the dynamics of (2.26) at large. Figure 2.27 illustrates scenarios for the destruction of single-frequency oscillations and the onset of beats (which occurs for sufficiently small f_0). We see that the limit cycle is born by the separatrix of the saddle equilibrium state when it merges with the stable equilibrium. Since the time of the motion along a closed trajectory is infinite, the frequency of the beats at the instant of their birth will almost equal zero, while their amplitude will be finite straight away – a limit cycle born from the separatrix has finite size. We may say that for weak signals the beats arise frequency-wise "softly" and amplitude-wise "hardly". In this case the shape of the beats is strongly nonlinear, similar to a cnoidal wave.[21]

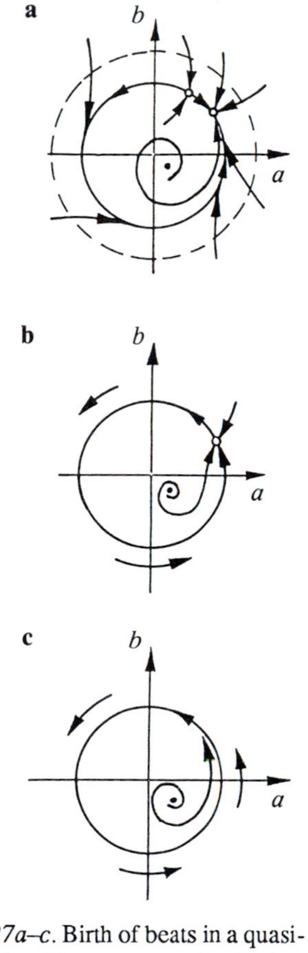

Fig. 2.27a–c. Birth of beats in a quasi-harmonic self-generator under the influence of weak external harmonic forcing. For increased detuning a limit cycle emerges from the separatrix loop.
a) Small detuning;
b) transition beyond the critical detuning value;
c) going beyond the synchronization interval

[21] Remember that the soliton corresponds also to a closed separatrix loop in phase space.

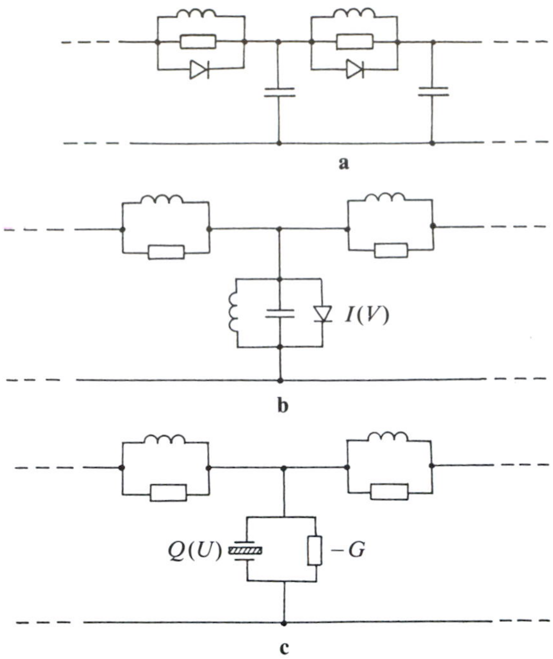

Fig. 2.28a–c. Examples of chains of generators:
a) chain with high-frequency instability;
b) discrete version of an active medium modelled by the Ginzburg-Landau equation;
c) chain with conservative nonlinearity and low-frequency instability

2.3.3 Self-Excited Oscillations in Chains and Continuous Systems

Having acquired some experience and intuition in the investigation of the sequence of nonlinear oscillators:
oscillator → coupled oscillators → chains → nonlinear conservative systems,
we can now follow the same route to study self-excited oscillations. We can consider self-excited oscillators both as an autonomous engineering system (generator networks, arrays of astrophysical antenna with controlled directivity pattern, etc.) or like a discrete model of a nonequilibrium medium in which oscillatory instabilities are stabilized by dissipative mechanisms. Examples of such chains of self-excited oscillators are given in Fig. 2.28.

It is obvious that the processes in dissipative media will differ from those in conservative media. Obvious deviations are time irreversability, memory loss with regard to the initial conditions, the emergence of an attractor corresponding to a constant level of pulsation intensity, etc.). The more remarkable is the generality of the ideas and approaches that enable us to distinguish "quasi-linear" and "strongly nonlinear" phenomena. The Fermi-Pasta-Ulam chain and the KdV have proved that even with a very weak nonlinearity strongly nonlinear phenomena may be observed. They manifest themselves through soliton solutions. This occurs in systems (media) with small dispersion. The same holds for a chain of linearly coupled quasi-harmonic oscillators (Fig. 2.28c) or in its continuum analogue. In spite of a weak nonlinearity (an individual oscillator yields a sinusoidal solution), collective self-excited oscillations in a chain are strongly nonlinear and, in particular, may yield a sequence of solitons, Fig. 2.29. This is readily understood. In such a system (like a ring, or a ring resonator) self-excited oscillations in the form of stationary waves are described by

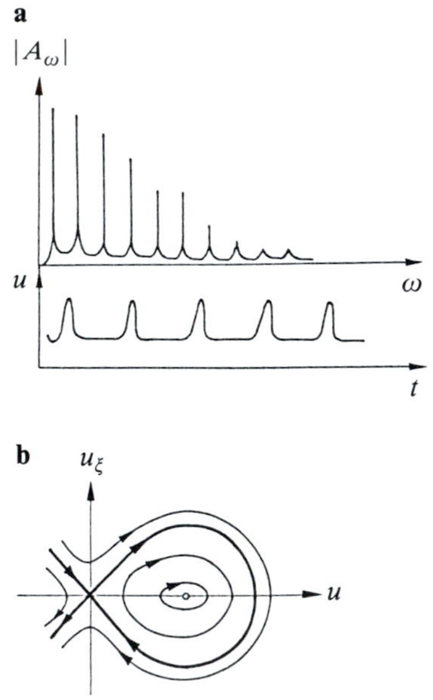

Fig. 2.29a,b. Periodic train of solitons observed
a) in an LC-chain with tunnel diodes [2.51];
b) the corresponding phase portrait for stationary waves

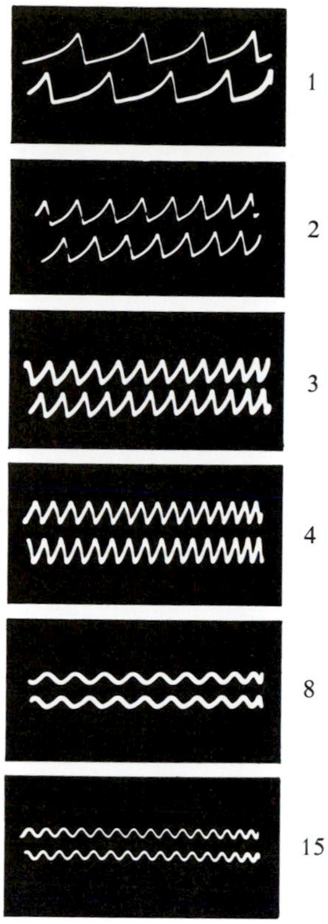

Fig. 2.30. Self-excited oscillations: a stationary running wave observed in a ring transmission line (Fig. 2.28a and [2.52]). The numbers in the figure denote different modes with the wavelengths $\lambda = L/2n\pi$ where L is the length of the ring

$$\beta\frac{d^3u}{d\xi^3} + \nu\frac{d^2u}{d\xi^2} + (V-1)\frac{du}{d\xi} + \gamma u(1-u^2) = 0 \qquad (2.27)$$

with $\xi = x - Vt$. Here β describes the dispersion and ν stands for high-frequency linear losses (imaginary dispersion). When $\beta \to 0$ and $V = 1$, all stationary waves periodic in space are described by closed trajectories in the phase plane $(du/d\xi, u)$; see Fig. 2.29b. This is the phase plane of a nonlinear oscillator! For a sufficient length of the ring (many oscillators in the chain, i.e. $N \gg 1$), a long periodic nonlinear wave is established. It corresponds to the closed trajectory in the vicinity of the separatrix. Such periodic sequences (of solitons) were also observed experimentally [2.51].

Not all one-dimensional self-excited oscillating chains with stationary waves are described by the oscillator equation. For example, the stationary wave equation (obtained in the continuum limit) of the chain specified in Fig. 2.28a has the more correct "self-excited oscillatory" form (compare with (2.22))

$$\nu\frac{d^2u}{d\xi^2} + \left[(V-1) + \alpha u^2\right]\frac{du}{d\xi} + \gamma u = 0\ . \qquad (2.28)$$

For $V < 1$ this is the simple Van der Pol equation whose phase plane has a single stable limit cycle. In this case, again we have individual generators that are close to harmonic ones. However, stationary waves having a large spatial period (again $N \gg 1$) may be strongly nonlinear and even steep, see Fig. 2.30. It is obvious that in this chain the short waves (for which only a few oscillators "work" simultaneously) will remain harmonic. Indeed, this is confirmed by Fig. 2.30 [2.51, 52]. Self-excited oscillations in multi-dimensional media are extremely versatile and often give rise to an intricate spatial pattern. Periodic self-excited oscillations in a cell with an electrolyte (aqueous solution of copper sulphate) are shown in Fig. 2.31. These self-excited oscillations [2.53] are due to the time-periodic "relocking" of four vortices giving rise to the observed variations of the flow patterns.

A regular self-excited oscillation is one of the simplest phenomena observed in nonequilibrium nonlinear media. The onset of structures having complex spatial organization and the appearance of chaos and turbulence are typical for such media. These problems will be discussed below. We shall come back to the classical results of the nonlinear theory of oscillations that have played an exceptional role in the development of modern nonlinear dynamics.

One of the most important achievements of the classical theory is the idea of the "historical" or "embryological" approach (Andronov) to the investigation of dynamic systems. Such an investigation proposes the analysis of the evolution of the phase space structure for varying parameters of the system. To make our further acquaintance with nonlinear physics more efficient we shall now attempt to put together all data available on the evolutionary behaviour of finite-dimensional systems.

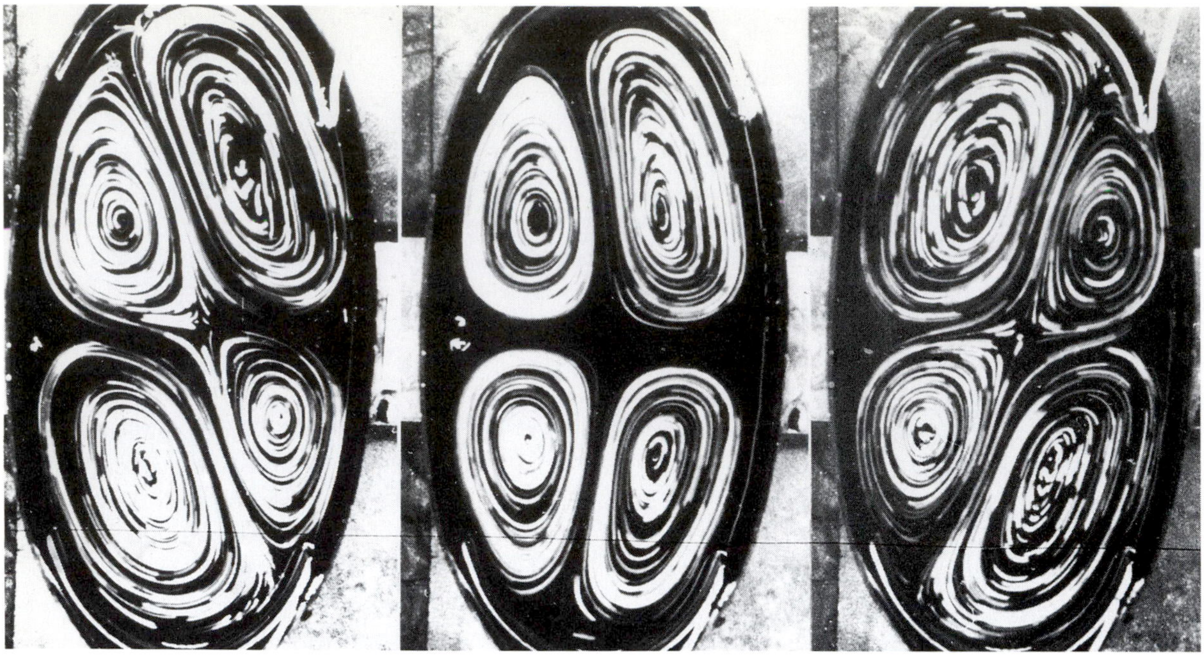

Fig. 2.31. Periodic self-excited oscillations in a cuvette with an electrolyte [2.53]. From *right* to *left*:
t_1 ;
$t_2 = t_1 + T/4$;
$t_3 = t_1 + T/2$

2.4 Bifurcations

We are well acquainted with gradual changes in the final results when the external action (e.g. force applied to the system) changes slightly. At first sight bifurcations seem to be something new and unusual since they imply that a minute change in the action may lead to a dramatically different result. But just try to cross a ridge by walking along a small board or even on a rope. Or think about examinations, lesser and lesser efforts in the preparation will reduce the number of points/credits to be earned until a bifurcation is reached with the two qualitatively different alternatives "pass" or "fail".

2.4.1 Acquisition of a New Quality

We have already referred to "bifurcations" and to results of the theory of bifurcations. Bifurcations figure prominently in the classical theory of oscillations and in modern nonlinear dynamics. The term "bifurcations" (from the latin word "bifurcus" standing for "forked", "separated into two part") means the acquisition of a new quality by the motion of a dynamic system with small changes in its parameters. Bifurcation corresponds to the restructuring of the motion of a real system (physical, chemical, etc.). The foundation for the theory of bifurcations was laid by A. Poincaré and A.M. Lyapunov back in the beginning of the 1920s and later on this theory was advanced further by A.A. Andronov, V.I. Arnold and others. Knowledge of basic bifurcations is a useful tool in the investigation of specific physical systems like the prediction of the parameters of new motions occuring during transitions, the evaluation of the region of their existence and

Bifurcation.
What's that?
Try to balance on a rope and you will know.

stability in parameter space, and so on. This refers both to discrete and continuous systems.

Mathematically, bifurcation is the change of the topological structure of partitioning the phase space of a dynamic system into trajectories with small variations of the system's parameters. This definition is based on the concept of the topological equivalence of dynamic systems. Two systems are topologically equivalent, i.e. have identical structures of the phase space partitioning into trajectories, if the motions of one of them can be transformed into the motions of the other via a smooth substitution of coordinates and time. An example of such an equivalence is the motion of a pendulum with different values of the friction coefficient: when the friction is small, the trajectories on the phase plane have the form of a twisted spiral and when it is large the trajectories are parabolas (Fig. 2.32). These – at first sight different – phase portraits can be transformed one into the other by introducing a new coordinate system, i.e. the transition from the phase portrait in Fig. 2.32a to that in Fig. 2.32b is not a bifurcation, because the bifurcation is the transition from a given system to a topologically nonequivalent one.

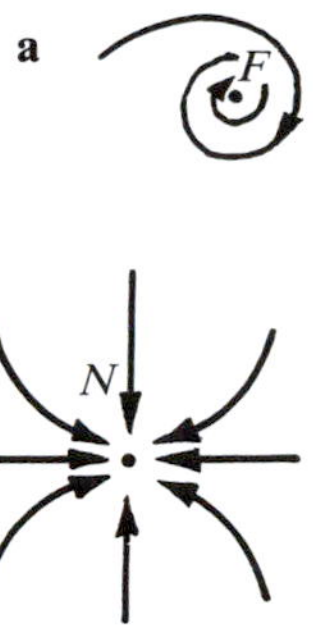

Fig. 2.32a,b. Phase portraits of the system $\ddot{x} + k\dot{x} + x = 0$ for
a) $k < 2$;
b) $k > 2$

An illustrative example of restructuring of motion of a real system is thermal convection in a horizontal layer of fluid heated from below: the increase in the temperature of the lower surface T_+ up to a certain temperature difference $(T_+ - T_-)$ does not cause macroscopic motion of the fluid (the heat flux between the lower and the upper surfaces is due to the molecular heat transfer); when the difference $(T_+ - T_-)$ reaches a critical value T_{cr} cellular convection arises, Fig. 2.33a. In the mathematical model (the initial equations of hydrodynamics in their finite-dimensional approximation) the appearance of such cells corresponds to the bifurcation of the birth of new equilibrium states, see Fig. 2.33b.

Among the great variety of bifurcations that are encountered in the analysis of models of physical systems the so-called local bifurcations are most interesting. They correspond to the restructuring of some motions of a dynamic system. Most simple and important of them are the bifurcations of equilibrium states and the bifurcations of periodic motions. They can be divided into three groups: the bifurcations of equilibrium states; the bifurcations of the birth of periodic motion (Table 2.1) and the bifurcations of the change of stability of periodic motions (Table 2.2).

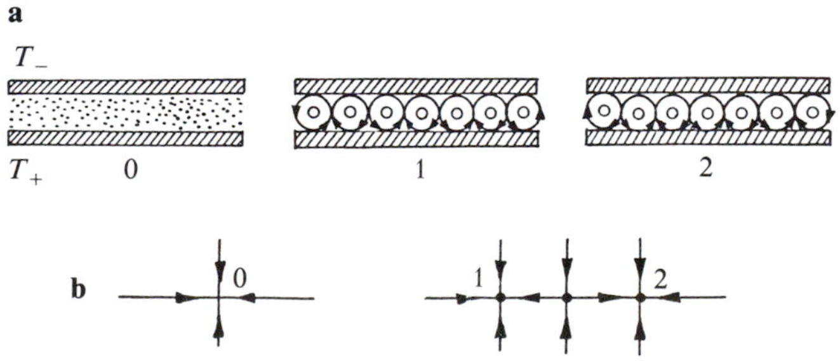

Fig. 2.33a,b. As convection becomes active, one observes the evolution from a single equilibrium state into three states (two stable ones and a trivial nonstable one); T_- and T_+ stand for the temperatures at the upper and lower boundaries of the horizontal liquid layer, respectively.
a) In real space: *(0)* no convection; *(1)* "left" convection; *(2)* "right" convection;
b) like *a)* but in phase space

Table 2.1. Birth of periodic motions. The second column specifies the type or character of the arising periodic motions (self-excited oscillations); the next column makes reference to equations modelling such systems; the following column gives additional comments while the sketches illustrate the phase portraits (from the *left* to the *right*) *before* bifurcation, *at the very moment* of bifurcation and *after* bifurcation, respectively

No.	Type of motion	Model	Comments	
1.	With regard to amplitude *hard* and *soft* with regard to frequency	Equations for the amplitude; van der Pol generator under the influence of periodic forces $\dot{a} = a[1-(a^2+b^2)] - \Delta\omega b - a_{\text{ext}}$ $\dot{b} = b[1-(a^2+b^2)] - \Delta\omega a$ $\Delta\omega$ – frequency detuning	In terms of the initial (not averaged) equations $\ddot{x} - \mu(1-x^2)\dot{x} + x = A\sin\theta$ $\theta = \omega$ this bifurcation depicts the birth of a torus. Experimentally that corresponds to the transition of a non-autonomous oscillator from synchronization to beats.	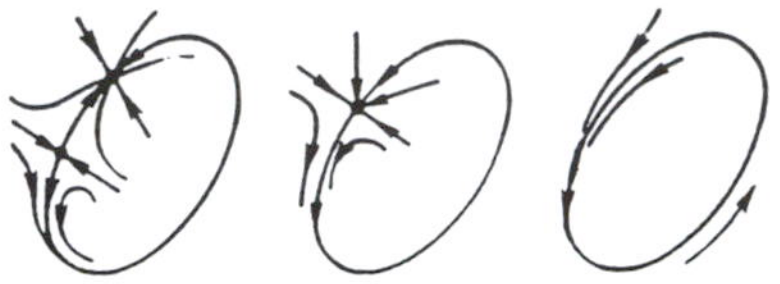
2.	As item 1.	Van der Pol - Duffing equation $\ddot{x} = \mu(1-x^2)\dot{x} + x - x^3 = 0$	For stationary waves in non-equilibrium media such a bifurcation corresponds to the transition from a quasi-harmonic wave to a soliton and then to a cnoidal wave	
3.	*Hard* with regard to amplitude and frequency	Equation of a self-excited generator with hard excitation $\ddot{x} + \mu(1 - x^2 + \alpha x^4)\dot{x} + x = 0$	This is one of the most typical bifurcations for the emergence or dissapearance of periodic motions	
4.	*Soft* with regard to amplitude and *hard* with regard to frequency	Van der Pol equation $\ddot{x} - (\alpha - x^2)\dot{x} + x = 0$	The Arnold-Hopf bifurcation is met in the most different branches of sciences	
5.	*Soft* with regard to amplitude and frequency		Such a bifurcation is realized when varying two or more parameters. Situations of this type arise in the equations of hydrodynamics	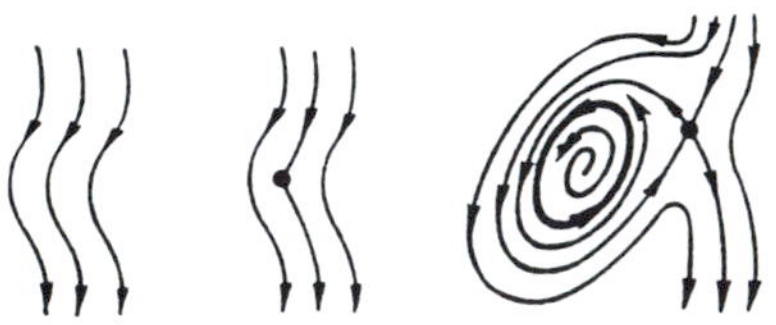

Table 2.2. Bifurcation in the change of the stability of periodic motions. The second column specifies the type of bifurcation/motion; the next column refers to equations modelling such systems; the following column gives additional comments. The illustrations show (from the *left* to the *right*) the respective phase portraits before and after bifurcation, and the respective multiplicators (with $\lambda = -1$; $\lambda = \exp[i\alpha]$ with $\alpha \neq \pi n/q$ and $\lambda = +1$ from the *top* to the *bottom*, respectively)

No.	Type	Model	Comments	
1.	Bifurcation of period doubling	Nonlinear oscillator with parametrically excited periodic force, e.g., $\ddot{x} - k\dot{x} + (1 + b\cos\theta)x + x^3 = 0$ with $\dot{\theta} = \omega$	An infinite chain of bifurcations of period doubling is one of the most general paths towards stochasticity in realistic systems	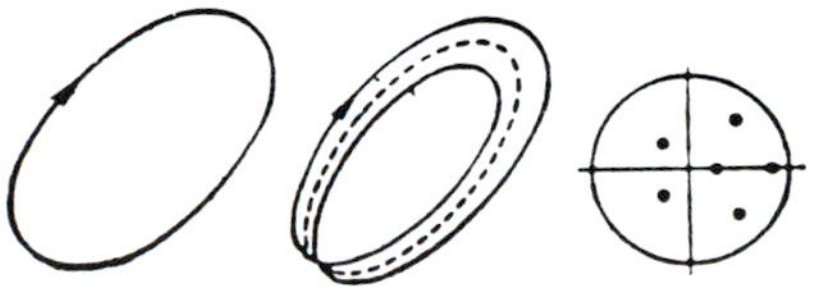
2.	Birth of two-frequency oscillations	Van der Pol generator under the influence of external forces $\ddot{x} - \mu(1 - x^2)\dot{x} + x = A\sin\theta$ with $\dot{\theta} = \omega$	For $\alpha = \pi n/q$ (with integers n and q and $\alpha \neq 0, \pi, 2\pi/3, \pi/2$) a torus emerges with stable and unstable periodic motions. For $\alpha = 0, \pi$, $2\pi/3, \pi/2$ no smooth torus arises and the situation is more complicated	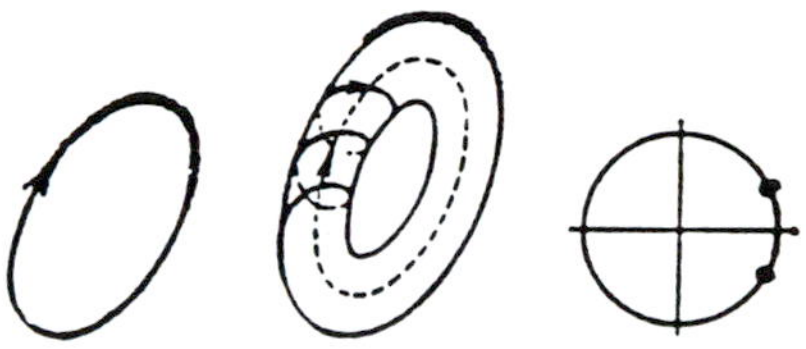
3.	Birth of a pair of stable periodic motions	Forced oscillations of an elastic rod under the influence of small periodic forces	Such a bifurcation is typical for nonlinear systems (with the potential energy exhibiting two minima) under the influence of external forces	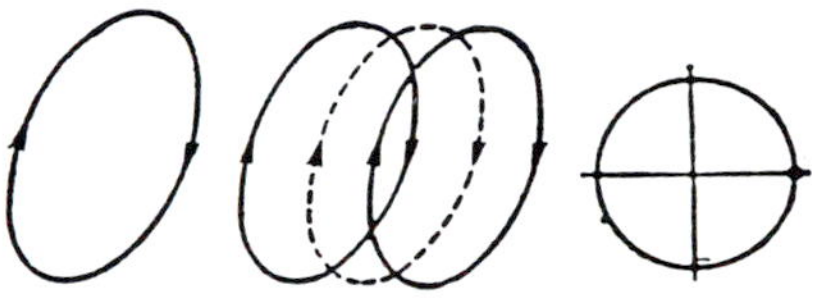

2.4.2 Bifurcations of Equilibrium States

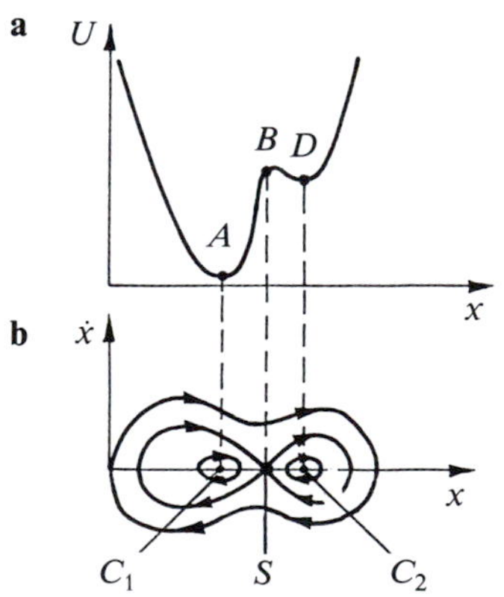

Fig. 2.34a,b. Motion of a marble in a potential well with a plateau
a) the potential U(x);
b) the phase portrait of the marble's motion

The principal bifurcations of equilibrium states are:

1) The merging and subsequent vanishing of two equilibrium states: An example is the motion of a marble in a potential well with a "plateau", see Fig. 2.34. When the plateau is smooth the equilibrium states, saddle S and center C_1, merge and vanish, Fig. 2.35.
2) The birth of a limit cycle from equilibrium states: An example of this bifurcation is the transition of a simple generator from a static state to a self-excited oscillatory regime with the change of the control voltage (Table 2.1). In this case, the stable focus at the origin of the coordinate system gives (in the phase plane $(u, \dot{u})$ for the damping constant $\alpha \geq 1$) birth to a limit cycle (Table 2.1, line 4) whose amplitude is for small α of the order of $\sqrt{\alpha}$ while the focus becomes unstable.
3) The birth of three equilibrium states from a single one (in theoretical physics such a bifurcation is referred to as spontaneous

symmetry breaking). For example, the change in the motion of the marble in a chute with a hump at its bottom corresponds to a bifurcation in which the degenerate equilibrium state of the center type (Fig. 2.36a) gives birth to three equilibrium states: the saddle S and two centers C_1 and C_2 (Fig. 2.36b), which provides conditions for the existence of stable nonsymmetric motions in a completely symmetric system.[22]

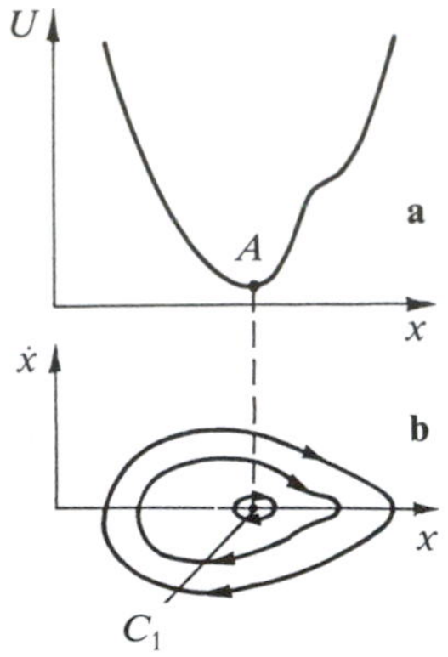

Fig. 2.35a,b. The marble after bifurcation;
a) the potential U(x);
b) the phase portrait of the motion

Let us observe the local bifurcations in the evolution of small perturbations in the system described by linearized equations. Thus, in the dynamic system $\dot{\boldsymbol{x}} = X(\boldsymbol{x}, \mu)$ (where $\boldsymbol{x}$ is the vector of physical variables, μ a parameter and $\boldsymbol{x}_0(\mu)$ the equilibrium state) the small perturbations $\boldsymbol{\xi}$ are described by the equation $d\boldsymbol{\xi}/dt = A(\mu)\boldsymbol{\xi}$ where $A(\mu) = \partial X[\boldsymbol{x}_0(\mu), \mu]/\partial \boldsymbol{x}$. If the roots λ_n of the characteristic equation $\det|A(\mu) - \lambda E| = 0$ (where E is the unit matrix) do not lie on the imaginary axis of the complex plane, no bifurcation takes place; only when for μ equal to its critical value μ_{cr} one or several roots lie on the imaginary axis of the complex plane, see Fig. 2.37.

An important thing is to be emphasized here: all bifurcations of the disappearance or birth of equilibrium states correspond to the passage of one or several roots through zero. An example is given in Fig. 2.38 where the birth of equilibrium state of the saddle S and node N types is presented. Such a bifurcation is encountered in the problem of mode competition in a self-generator and, for example, in the problem of competition of species with the numbers x_1, x_2 feeding from one source. The corresponding kinetic equations describing the variation of numbers have the form

$$\dot{x}_{1,2} = [1 - (x_{1,2} + \varrho_{2,1} x_{2,1})]x_{1,2} , \tag{2.29}$$

see (2.25). When $\varrho_{1,2} > 1$, any species may win in the struggle for existence. As one of the parameters ϱ_1, ϱ_2 is decreasing and becomes smaller than unity, only a single one of these species will survive under arbitrary initial conditions, see Figs. 2.26a,c. Similar situations describe the competition of various spatial structures appearing in fluids with thermal convection and in other situations characterized by multi-stability, see Chap. 4, *Structures*.

When two roots of a characteristic equation become purely imaginary, then from an equilibrium state either a limit cycle is born or dies (Table 2.1, line 4). This means that for all values of the parameter μ that are smaller (or larger) than the critical μ_{cr} or rather close to it, there exists a periodic solution that tends for $\mu = \mu_{cr}$ to the static form $\boldsymbol{x}_0(\mu)$. The stability of the limit cycle then inherits the stability of the equilibrium state at the precritical parameter value. Such a bifurcation is known as the Andronov-Hopf bifurcation [2.54].

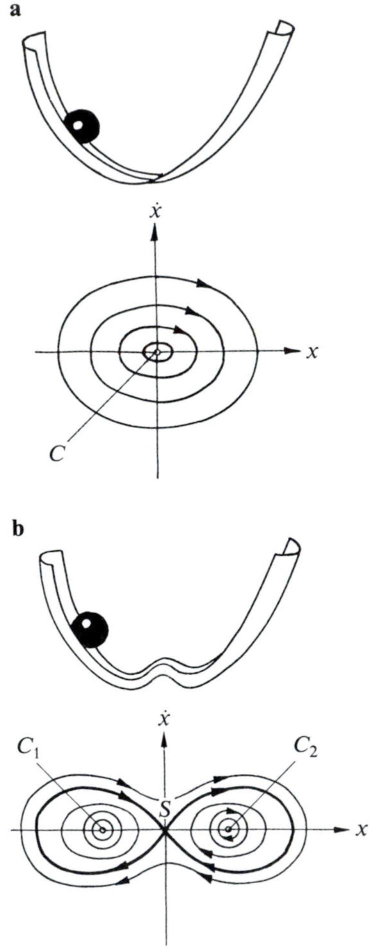

Fig. 2.36a,b. Birth of three equilibrium states out of a single state:
a) single state;
b) the emerging three equilibrium states

[22] Note that the symmetry of this system by no means implies the symmetry of motions (solutions) that are realized. Such a symmetry guarantees only equal probability of solutions with opposite types of "non-symmetry".

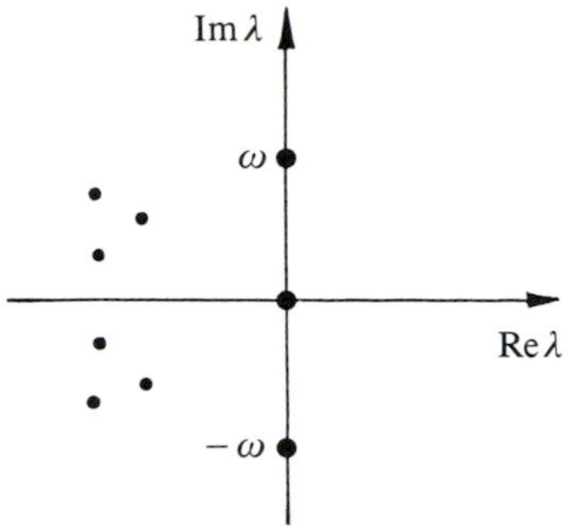

Fig. 2.37. Location of the roots $\mu = \mu_{cr}$ in the complex plane

2.4.3 Bifurcations of Periodic Motion

Table 2.1 summarizes the basic types of the bifurcations of the birth (when the phase portraits are viewed from left to right) or vanishing (if viewed from right to left) of periodic motions. They are classified into three groups. If we take the vanishing of periodic motions, the first group (the two first lines in Table 2.1) includes the bifurcations in which the time T of periodic motion tends to infinity (or the frequency ω vanishes) as $\mu - \mu^* \to 0$, while the average oscillation amplitude is different from zero. An example of such a bifurcation in self-excited oscillating systems is the onset of modulation under the action of periodic force on the self-generator. The limit cycle – the image of modulated oscillations – is born from a separatrix "saddle-node" loop as two equilibrium states: a saddle and a node, merge and disappear (Table 2.1, line 1). Knowing such a bifurcation one can determine the properties of the regime emerging after the transition through the critical point: the established modulation will have finite amplitude and modulation frequency close to zero.

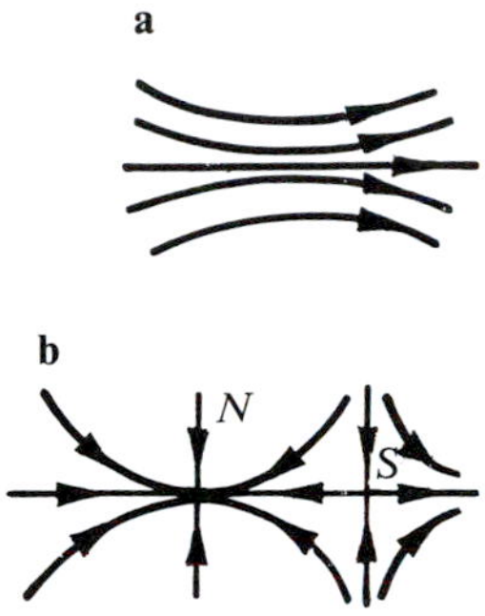

Fig. 2.38. Smooth phase flow (*a*) giving rise to an equilibrium state (*b*)

The second group is represented by the bifurcation of the disappearance of stable periodic motion at the instant of its merging with unstable periodic motion (Table 2.1, line 3) which is referred to as a tangential bifurcation. This bifurcation is typical for a self-excited oscillator with hard excitation, its Poincaré mapping is given in Fig. 2.39. Figure 2.39a depicts the state of the system in the absence of stable oscillations, i.e. there are no limit cycles. Fig. 2.39b corresponds to the moment of bifurcation: the curve $x_{n+1} = f(x_n)$ is tangent to the bisectrix of the first quadrant – two periodic motions, stable *1* and unstable *2*, are born, see Fig. 2.39c.

The bifurcations of the third group are, as a rule, encountered in systems depending on two or more parameters (Table 2.1, line 5).

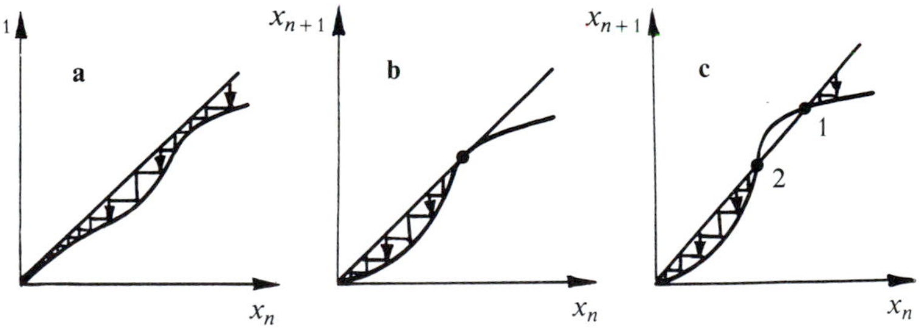

Fig. 2.39a–c. Birth (vanishing) of a stable threshold cycle (fixpoint) on a Poincaré map:
a) no stable cycle;
b) bifurcation with a halfstable cycle;
c) stable (*1*) and unstable (*2*) cycle

2.4.4 Bifurcations – Changes of Stability in Periodic Motion

An important feature of the bifurcations of the change of stability in periodic motions (Table 2.2) is the value of multipliers at a critical moment. Physically, multipliers are the amplification or damping constants of small perturbations against the background of periodic motion in period T. Mathematically, multipliers are the eigenfunctions of the matrix $\exp[RT]$ describing the solution $\xi(t) = C(t)\exp[RT]$ of a linearized system in the neighbourhood of the periodic motion

$x = F(t, \mu)$ with $F(t + T, \mu) = F(t, \mu)$. R is a constant and $C(t)$ is the periodic matrix $C(t+T) = C(t)$. It should be noted that one of the multipliers in an autonomous system described by time-independent equations is always unity (there is, on the average, neither stretching nor contracting along the periodic trajectory). Therefore, below we shall consider only the other multipliers. If all other multipliers have absolute values less than unity, then the initial periodic motion will be stable. Bifurcations related to the loss of stability occur at the parameter values of the system for which one or several multipliers have absolute values equal to unity. Thus, the multipliers lie inside a unit circle on the complex plane, see Table 2.2.

... Sometimes a small difference in the intitial states can be the cause of a considerable difference in the final phenomenon. A small error in the first would bring about a high error in the latter. The prediction becomes impossible, we have an accidental phenomenon.

H. Poincaré

If one of the multipliers equals -1, the well-known (see Chap. 3 *Chaos*) bifurcation of period doubling occurs (Table 2.2, line 1). During the bifurcation a small absolute perturbation simply changes sign over one period, then on the next orbit the disturbed trajectory will be closed in the linear approximation. A stable periodic motion with twice the period is born from the initial periodic motion and the original motion will be unstable.

The bifurcation of the birth of a two-dimensional torus from the periodic trajectory (Table 2.2, line 3) corresponds to the appearance of bifrequency, quasi-periodic oscillations in a physical system, as A. A. Andronov metaphorically put it "the cycle loses its skin" in this bifurcation. Bifurcations in which two stable limit cycles (Table 2.2, line 3) are born simultaneously are encountered in systems depending on two parameters or in systems with a specific type of symmetry.

Bifurcations resulting in the vanishing of static or periodic regimes (i.e. equilibrium states or limit cycles) may cause the dynamic system to enter a regime of stochastic oscillations (see Chaps.3 and 5).[23]

2.5 Modulation

Modulation is the process that enables us to transmit not just simple signals, but even highly sophisticated ones, music and pictures. Radio and TV would not be possible without a knowledge of modulation and a mastering of the associated technology.

2.5.1 The Role of Small Parameters

Modulation is an adiabatic variation of the parameter of oscillations, waves, structures, etc. An alternative process is "catastrophes" or "bifurcations", i.e. abrupt variation of the objects or their features. Slow variation in the parameters of oscillations or waves implies that the forces causing these changes are also small. In the basic equations, this is expressed through a small parameter that provides an opportunity to

[23] Note that the term "bifurcation" is sometimes used to denote the restructuring of objects which do not change in time; in this case the term "catastrophe" is also employed (see, e.g. [2.55]).

describe modulation phenomena using an asymptotic method. However, we must not jump to the conclusion that most effects associated with modulation can be described within "quasi-linear physics".

Let us recall, for example, the problem of the synchronization of a generator by a harmonic external force at weak signals. The basic equation of the problem,

$$\ddot{x} + \omega_0^2 x = \mu \left[\left(1 - x^2\right) \dot{x} + f_0 \sin \omega t \right] ,$$

contains the small parameter $\mu \ll 1$. The oscillations in the system are actually quasi-harmonic and the pulsations are low-frequency (slow) ones. But even averaged equations like (2.26) no longer contain a small parameter. At the same time, if we want to describe the birth of pulsations, it is not sufficient to analyse the behaviour of the system only near equilibrium, the behaviour of trajectories need to be investigated in a finite region of phase space. Thus, the quasi-linearity of the modulation problem can be employed to construct the approximate equations for slowly varying parameters, but the analyses of "modulation proper" will already encounter a strongly nonlinear problem.

Since nonlinear physics is concerned, as a rule, with modulation phenomena (without modulation the signal carries no information, in fact, there is no signal at all), it is important to recall the basic models and effects in the nonlinear theory of modulated oscillations and waves. Getting a little ahead we can say that all or nearly all effects encountered in original oscillations or fields are inherent in modulation. These are, in particular, the formation of solitons and shock waves, competition, synchronization, recurrence, dynamic chaos, etc. Let us make sure that it is really so.

A modulated oscillation or wave is by no means necessarily a "sinewave with slowly varying amplitude and frequency"; the shape of the elementary oscillation or filling wave on which the modulation is superimposed can be arbitrary within broad limits (for example, a periodic cnoidal or saw-tooth wave). We may speak with some justification of modulation as a slow change in certain parameters of motion even when the "filling" is not periodic. And while the analysis of modulated oscillations that are close to periodic non-sinusoidal oscillations has roots in classical oscillation theory, the investigation of slowly evolving nonperiodic oscillations or waves is characteristic of the modern theory. The first problem of this kind arose in the study of the behaviour of nonlinear waves in media with slowly varying parameters. Here we have a very pictorial example – the evolution of sea waves as they approach the shore. This problem was solved by analysing "quasi-solitons", i.e. waves similar to solitary stationary waves with amplitudes varying slowly due to the inhomogeneity of the medium (which arises because of the changing depth near the shore) [2.24].

The next important distinguishing feature of the modern theory is essentially its broadening of the modulation concept to include not only modulation transformations, but also modulation generation – self-modulation.

Modulation is used to convert a continuous train of waves into a carrier of information.

Like oscillations, modulation may arise as a result of instability (self-modulation), may be forced (modulation is transferred to the carrier from an external source) and, finally, may be specified at the initial time (an analogue of free oscillations). Nearly all phenomena typical for oscillations exist also for modulated waves. This will perhaps seem less surprising if we recall that in many cases the transformations that occur in the modulation spectrum of a nonlinear system differ from the corresponding spectral changes of the oscillations (or waves) themselves only in that they occur in a higher frequency range (are transplanted to the carrier frequency).

2.5.2 Running Mandestam Lattices. Modulation of Waves by Waves

Today, when we speak of modulation of waves by waves, the picture of a periodic traveling lattice on which the incident wave is diffracted (modulated) seems so natural that we do not concern ourselves with its origin. Mandelstam was the first to see this classical picture. As early as 1913, analysing the scattering of light at the interface between two media, he "materialized" the terms of the spatial Fourier series, independently of Einstein and Debye, placing real periodic lattices in correspondence to them (just as, a little earlier, he had indicated the reality of spectral satellites in time modulation of alternating current). But they were still stationary lattices. Traveling lattices appeared five years later. By that time, Debye's theory of the heat capacity of solids, in which elastic (acoustic) waves were presented as "storages" for energy of thermal motion, was quite well known, and Mandelstam was the first to point out that light scattered by thermal fluctuations should be frequency-modulated by a traveling acoustic wave (lattice) and found the frequencies of the satellites: $\nu_\pm$ with $(\nu_\pm - \nu) = \pm 2\pi\nu(C_{\text{sound}}/C)\sin(\theta/2)$, where ν is the frequency of the incident light, C_{sound} and C are the velocities of sound and light, and θ is the scattering angle. This was a prediction of the scattering of electromagnetic waves by acoustic waves (Mandelstam – Brillouin scattering)[24] – the first example of the scattering of waves by waves that is now widely investigated in many areas.

Ten years later (in 1928) L.I. Mandelstam, working in collaboration with G. S. Landsberg on the study of the scattering of light in crystals, attempted to observe spectral satellites produced by modulation of light by sound. However, they observed much stronger splitting, which they explained in terms of modulation of the light by infrared vibrations of molecules. They discovered the Raman scattering of light – the scattering of waves by oscillators.[25] The scattering of light by

M.A. Leontovich
1903–1981. Work related to electrodynamics, optics, thermodynamics, statistical physics, quantum mechanics, theory of oscillations, acoustics, quantum electronics, plasma physics, thermonuclear synthesis.

[24] By this time, L. Brillouin had already published some of his results on the scattering of light by sound.

[25] The optical branch of the dispersion curve had already been reported, but Mandelstam and his colleagues must have been unaware of those studies: exchange of information had not yet resumed after interruption by the war. They identified this branch independently (*M.A. Leontovich* [2.56]).

Ch.V. Raman
1888–1970. Awarded the 1930 Nobel Prize for phyics for his work on the scattering of light and for the discovery of the Raman effect.

R.L. Weber
ibid. p.93

sound was not observed experimentally until 1932 in France and in the United States.[26] The Mandelstam-Brillouin scattering (MBS) and Raman scattering (RS) are modulations as understood by Mandelstam: *In much the same way as you inject your speech into the emission of a radio station by means of modulation, so do atoms oscillating in a molecule or a crystal lattice tell us of their infrared vibration, using the frequency of the emitted light as a carrier*. This modulation of an incident wave by specified sources is usually called spontaneous scattering.

In modern nonlinear theory, attention is focussed on processes discovered in the early 1960s: induced scattering by oscillations (IRS, 1962) and by waves (IMBS, 1964) [2.58–60]. In induced scattering the incident wave itself amplifies the modulation sources – oscillations of atoms or molecules in RS or the sound wave in MBS. These processes are one of the manifestations of the parametric instability that we have already discussed, i.e. the decay of the incident wave into a resonant pair of waves or into a wave and an oscillation (in RS).

When waves interact, modulation may either be produced or transferred from one wave to another. The first effect of this kind was actually observed back in 1930, when Tellegen (Luxembourg) and Lbov (Nizhnii Novgorod) tuned receivers to the frequency of a local radio station and received the transmission (modulation) of a powerful station working on a totally different frequency – cross-modulation. The just mentioned effect is explained quite simply (1934 [2.61]): on passage of a strong modulated wave (pump) through a volume of ionospheric plasma, the absorption coefficient for a weak wave also passing through this volume is changed in accordance with the modulation law specified by the plasma. Thus, the pump modulation is transferred to a different carrier.

Modulation transfer is possible not only from a strong (pump) wave, but also from a weak (signal) wave in the presence of an unmodulated pump. One of these possibilities is known to be embodied in the superheterodyne receiver – the modulation is amplified, mainly, at the intermediate frequency. A similar "superheterodyne" process can also be brought about for waves in a nonlinear medium with "intermediate" frequency amplification [2.62]. The equation for the amplitudes of the parametrically coupled waves $\omega_3 = \omega_1 + \omega_0$ in such a medium at a given pump (heterodyne) field a_0 = const can be written in the form (a_3 is the amplitude of the signal):

$$\partial_x a_1 = i\sigma_1 a_0^* a_3 + \gamma a_1 \ , \quad \partial_x a_3 = i\sigma_3 a_0 a_1 \ . \tag{2.30a}$$

This mechanism is of interest, of course, only if the amplification of the intermediate wave is strong enough $\gamma \gg \Gamma = |a_0|\sqrt{\sigma_1\sigma_2}$. Here the evolution of the signal and intermediate waves along the "receiver" is described by the following solution of (2.29):

[26] Raman (combination) scattering was discovered by *Raman and Krishnan* [2.57] simultaneously with Mandelstam and Landsberg.

$$\begin{aligned} a_1(x,t) &= a_3(0,t)\frac{i\sigma_1 a_0^*}{\gamma}(e^{\gamma x} - 1) ; \\ a_3(x,t) &= a_3(0,t)(1 - \delta e^{\gamma x}) \end{aligned} \tag{2.30b}$$

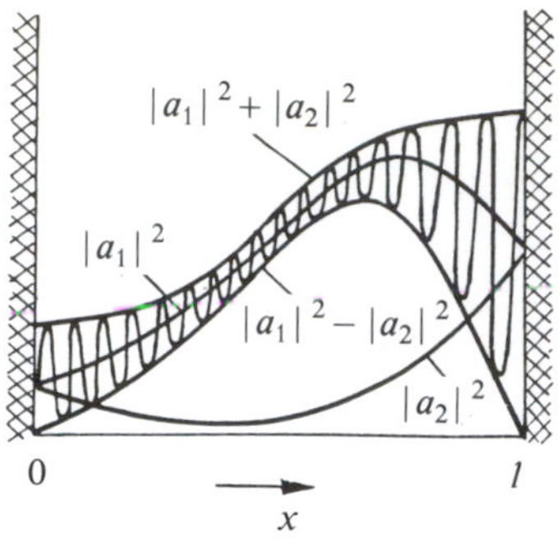

Fig. 2.40. Asymmetric spatially inhomogenous regime in a resonator with ideally reflecting walls (filled with a nonlinear nonequilibrium medium)

with $\delta = \Gamma^2/\gamma^2$. Where $a_3(0,t)$ is the modulated signal wave at the input ($x = 0$) into the nonlinear medium $a_2(0,t) = 0$. The process of signal amplification in a wave superheterodyne receiver can be described as follows. First, there is slight amplification of the intermediate wave a_1, onto which the modulation that existed at the input is transferred from the signal wave, in the interval $0 < x \lesssim 1/\gamma$; then the intermediate wave carrying the signal-wave modulation is strongly amplified in the range $1/\gamma \lesssim x \lesssim x_0 = \ln(1/\delta)/\gamma$, and, finally, the amplified modulation is transferred to the signal wave: $x > x_0$. This process is obviously also possible with a low frequency pump.

The waves may change the amplitudes of one another in a purely energetic interaction as well. In this case, we encounter quite peculiar effects of breaking the symmetry of solution with a complete symmetry of the problem. In particular, the field distribution that is not symmetric along the spatial coordinate, may be established in nonequilibrium dissipative media, in a fully isotropic medium and under completely symmetric boundary conditions. These nonsymmetric distributions of potentially equal counter-propagating waves are established, in particular, in a continuous generator with perfect reflection at the boundaries, see Fig. 2.40. This amazing phenomenon is explained by the effect of wave competition (but in space rather than in time). The equation for the amplitudes $a_{1,2}(x,t)$ of such waves for simple idealization is written in the form

$$\partial_t a_{1,2} \pm v\partial_x a_{1,2} = h\left[1 - \alpha\left(|a_{1,2}|^2 + 2|a_{2,1}|^2\right)\right] a_{1,2} \tag{2.31}$$

with boundary condition $|a_1(x,t)| = |a_2(x,t)|_{x=0,l}$, where l is the length of the resonator. The intensity distribution $|a_{1,2}(x)|^2$ in a stationary regime is readily reproduced from the form of the trajectory in the phase plane (2.26) for the partial derivative with regard to time being zero, see Fig. 2.41. In a short resonator, where the competition effect has not time enough to be manifested, only a banal regime of a standing wave is possible, with the corresponding equilibrium state on the curve $|a_1|^2 = |a_2|^2$ in the phase space shown in Fig. 2.41. While in a long resonator, the counter-propagating waves, drawing energy from a common source, suppress one another in the greater portion of the resonator and equalize only near reflecting walls. The resulting regime of standing wave is unstable and one of spatially inhomogeneous regimes is established, to which the trajectories of the ebc or bac type correspond (Fig. 2.41) – we again have a "strongly nonlinear" problem!

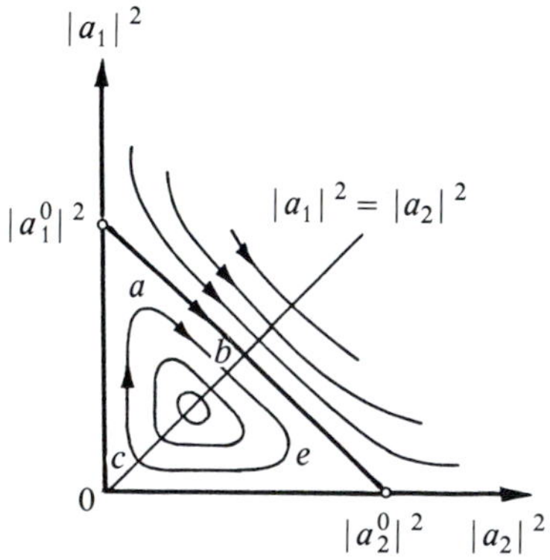

Fig. 2.41. Phase portrait of the system (2.31) at $\partial a_{1,2}/\partial t = 0$

The modulation of waves by waves is not always manifested in forms as familiar as slow variation of wave amplitudes or phases. Thus, when counter-propagating waves interact even in an isotropic nonlinear medium, their type of polarization may also change – the plane of linearly polarized waves may rotate, linear polarization may be converted into elliptical, and so on. Let us illustrate one of these effects with a specific example. We shall consider the interaction in time of spatially homogeneous counter-propagating waves of the same frequency in an optically active (nonequilibrium) medium. Let the angle between the field vectors of these linearly polarized waves be initially very small.What will then happen to them? The polarization vectors of counter-propagating waves will rotate in opposite directions relative to the direction of wave propagation, see Fig. 2.42 [2.63].

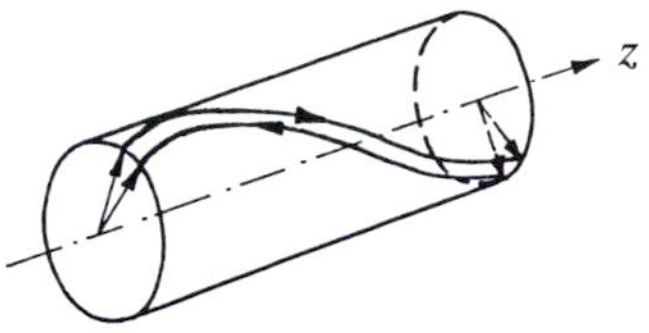

Fig. 2.42. Rotation of the polarization vector in the nonlinear interaction of counter-propagating waves

The polarization-rotation effect of counter-propagating waves was confirmed experimentally back in 1970 for a resonance isotropic active medium [2.64]. Now it has been observed in a wide variety of isotropic media. Since the magnitude of the effect – the angle of rotation or the ellipticity of the polarization of counter-propagating waves – depends in a subtle fashion on the properties of the nonlinear media, this effect proved to be useful for their diagnostics (nonlinear polarization spectroscopy) [2.65].

2.5.3 Generation of Modulation

The very first experiments on SMBS and SRS (or IMBS and IRS, respectively, see above) in optics revealed that the backscattered beam approximately repeats the evolution of the pump beam in the backward direction in time. Then it was found that in many experimental situations the scattered wave exactly reproduces a complex-conjugate incident wave that is strongly modulated in the transverse direction [2.66]. Duplication of the backscattered (Stokes) wave in the backward direction of the optical path traversed by the pump means that a limited region in which scattering occurs behaves like a mirror. But this is not an ordinary mirror: the reflected wave duplicates the optical path of the incident wave in forward time only where its phase front is conjugate with the pump, i.e. $a_p(\boldsymbol{r}) \sim a_0^*(\boldsymbol{r})$. The total phase of the wave $\exp[i(\omega t - kx + \phi)]$ in this case varies as it propagates in the x-direction in the same way as that of the incident wave in the backward time direction. This is why the effects of the reproduction of the transverse modulation of the pump beam in induced scattering radiation have come to be known as "phase conjugation" [2.67, 68].

The fact that the scattering volume acts as a nontrivial mirror is related to the selective manner of amplification of the Stokes wave (grown out of noise) in the field of a pump broken up in $\boldsymbol{r}$. If the phase front of the pump is unmodulated, Stokes waves with arbitrary transverse structure are amplified equally in its field; but if it is sufficiently cut up, the Stokes wave modulated in $\boldsymbol{r}$ such that its maxima fall onto the minima of the pump and vice versa is amplified worse than the one which duplicates the pump profile. Formally, this can be explained as follows: the total power (averaged across the beam) of the backscattered wave is described by the equation [2.67] $dP/dx = -g(x)P(x)$, where the gain in the direction of propagation is given by

$$g(x) = \frac{G \int a_0(\boldsymbol{r}) a_0^*(\boldsymbol{r}) a_p(\boldsymbol{r}) a_p^*(\boldsymbol{r}) d^2\boldsymbol{r}}{\int |a_p(\boldsymbol{r})|^2 d^2\boldsymbol{r}} \tag{2.32}$$

If, provided that $a_0(\boldsymbol{r})$ varies rapidly, the pump and initial noise intensities are uncorrelated in $\boldsymbol{r}$, the gain is $g = G < |a_0(\boldsymbol{r})|^2 >$ (quadruple correlations decay into pair correlations). But if $|a_p(\boldsymbol{r})|^2 \sim |a_0(\boldsymbol{r})|^2$, the increment will be twice as large. Since these two appear in the argument of the exponent and the total gain along x is sufficiently large, we can certainly say that from the backscattered noise background there will be extracted the wave with the phase conjugated front. Such effects are now widely used, in particular, for self-correcting transfer of powerful laser radiation over long distances – adaptive nonlinear optics [2.68].

2.5.4 Self-Modulation

Let us perform the following experiment: at the boundary of an LC-transmission line or chain of oscillators with a cubic nonlinearity (see (2.14)) we apply a sinusoidal oscillation whose frequency lies in the range of strong dispersion $\omega(k)$ (for example, at the bending of the dispersion curve shown in Fig. 2.16) and the resulting harmonics due to nonlinearity are out of synchronism with the main wave (and consequently, do not build up). What kind of oscillation will we observe at the other end of the line? The answer is found in oscillograms in Fig. 2.43: the oscillations are found to be modulated! It seems quite unexpected because we intuitively link the appearance of modulation (in the narrow sense of the word) only to the transfer of information on a low-frequency signal to a high-frequency carrier. We have already seen that the physical origin of this process may vary greatly, but there must be some source of modulation! But this isn't evident in our experiment. This example illustrates the phenomenon of self-modulation – modulation occurs as a result of parametric instability developing

a

b

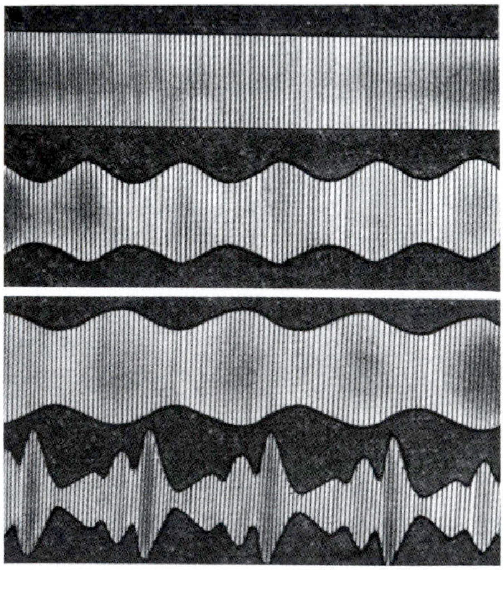

Fig. 2.43a,b. Self-modulation in a nonlinear transmission line:
a) onset of modulation;
b) evolution of a sinusoidal modulation wave

along the line, which, in this case, results in the appearance of satellite waves with frequencies ω_1 and ω_2 close to ω_0, where $\omega_1 + \omega_2 = 2\omega_0$ (this process is often referred to as the decay of a pair of quanta in the same state). This version of parametric instability is called modulation instability in the theory of nonlinear waves [2.69, 70].

To describe this and related phenomena in more detail, we shall have to consider the basic equation of the theory of modulated waves in nonlinear media, i.e. the nonlinear parabolic equation, or the nonlinear Schrödinger equation [2.71]:

$$(\partial_t a + v\partial_x a) - \frac{i}{2}\frac{d^2\omega}{dk^2}\partial_{xx}a - \frac{i}{2k}\Delta_\perp a = i\varepsilon_H\left(|a|^2\right)a \;. \tag{2.33}$$

Here a is the complex amplitude of the wave $\exp[-i(\omega t - \boldsymbol{kr})]$; its wave vector is $\boldsymbol{k}$ and the parameter ε_H characterizes the degree of nonlinearity of the medium; for example, for light waves $\sqrt{\varepsilon_H}$ is the nonlinear correction of the refractive index. For a simpler case of plane waves, instead of (2.33) we can write

$$(\partial_t a + v\partial_x a) - \frac{i}{2}\frac{d^2\omega}{dk^2}\partial_{xx}a + i\alpha a|a|^2 = 0 \;. \tag{2.34}$$

No modulation – no radio, no TV.

am – amplitude modulation
fm – frequency modulation

The term in the parentheses describes modulation waves traveling with the group velocity in a linear medium without dispersion; the parabolic term proportional to $d^2\omega/dk^2$ is responsible for dispersion spreading, and α for the magnitude and sign of nonlinearity.[27] We shall now see, modulation instability is possible only with a certain relation between the signs of the nonlinearity and the dispersion of group velocity, $\alpha d^2\omega/dk^2 < 0$ [2.71]. The physical mechanism of this limitation (which is usually referred to as the Lighthill condition) is most easily understood if we consider self-modulation not in space-time language, i.e. not from the analysis of (2.34), but in spectral language, restricting the analysis to the interaction-oscillators of only three waves that form a wave with sinusoidal modulation.

Equations similar to (2.5) are derived from (2.32) for the complex amplitudes of the satellites $\omega_\pm$ and the carrier ω_0:

$$\begin{aligned} &\dot{a}_0 + i\frac{\alpha}{2}|a_0|^2 a_0 = 0 \;; \\ &\dot{a}_\pm + i\left(\alpha|a_0|^2 + \frac{1}{4}\frac{d^2\omega}{dk^2}k^2\right)a_\pm = -i\alpha a_0^2 a_\pm^* \;. \end{aligned} \tag{2.35}$$

Here it has been taken into account that because of the spectral proximity of the satellites, the detuning is

$$\delta = 2\omega_0 - \omega(k_0 + k) - \omega(k_0 - k) \approx \left(d^2\omega/dk^2\right)k^2 \;.$$

The parametric increment γ with which the amplitude of the satellites increases in a given carrier field equals

[27] The parabolic equation describing beam diffraction was first derived by M.A. Leontovich in 1944.

$$\mathrm{Re}\{\gamma\} = \pm k \left[-\frac{d^2\omega}{dk^2}\alpha|a_0|^2 - \frac{k^2}{4}\left(\frac{d^2\omega}{dk^2}\right)^2 \right]^{1/2} . \tag{2.36}$$

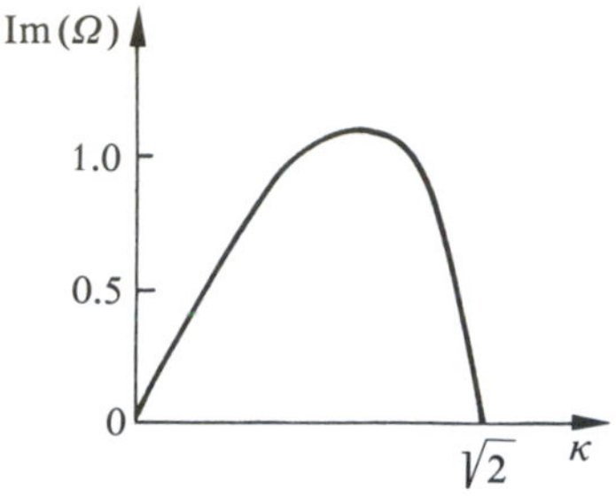

Fig. 2.44. Dependence of the increment Im(Ω) on the scale of modulation κ

Since the spatial scale of modulation may be arbitrary, a necessary (and for $k \to 0$ also sufficient) condition for modulation instability is $\alpha\omega_{kk} < 0$. Its physical meaning is now also clear: for the instability to appear, the nonlinear detuning $\sim \alpha|a_0|^2$ has to compensate the linear detuning $\sim (d^2\omega/dk^2)k^2$. Figure 2.44 shows the dependence of the increment on the modulation scale; for short-wave modulation, when $\Lambda^2 < (\alpha|a_0|^2/\pi^2)(d^2\omega/dk^2)$, the nonlinear detuning is no longer capable of compensating the dispersion spreading and the modulation does not become deeper (the increment becomes imaginary). The self-modulation effect was predicted in 1965 [2.69] and was observed experimentally a year later for waves on the surface of a liquid [2.72]. This effect is, presumably, [2.69] related to the explanation of the "tenth wave" phenomenon.

Now let us come back to the analysis of the evolution of modulation waves within the framework of the linearized equation (2.33). For these waves we obtain the dispersion law

$$\begin{aligned}\Omega(k, k_\perp) =& vk \pm \left[\left(\frac{v}{2k_0}k_\perp^2 + \frac{1}{2}\frac{d^2\omega}{dk^2}k^2\right)\right. \\ &\times \left.\left(2\alpha|a_0|^2 + \frac{v}{2k_0}k_\perp^2 + \frac{1}{2}\frac{d^2\omega}{dk^2}k^2\right)\right]^{1/2}\end{aligned} \tag{2.37}$$

which yields directly, in particular for one-dimensional waves $k_\perp \equiv 0$, the known modulation-instability increment (2.36). And what will happen within the framework of our basic model (2.33) to small multi-dimensional disturbances? Assuming for simplicity in (2.37) that $k = 0$, we find that at $k_\perp^2 < 2\alpha|a_0|^2 k_0/v$ the quantity $\Omega(k_\perp)$ is purely imaginary – the multi-dimensional perturbations with the frequency equal to the filling frequency will grow! Physically, this is manifested in the following way. If a plane wave of frequency ω_0 is incident on the boundary of a nonlinear medium whose dielectric constant or permittivity increases with field strength, the wave is transformed to a periodic (in the transverse direction) system of beams, i.e. it is self-focussed [2.73]. This is a stationary spatial variant of parametric instability or decay of a pair of quanta in the same state $2\boldsymbol{k}_0 \to \boldsymbol{k}_1 + \boldsymbol{k}_2 + \Delta\boldsymbol{k}(|a_0|^2)$, see Fig. 2.45.

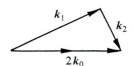

Fig. 2.45. Decay of a single state into two quanta

2.5.5 Recurrence

The nonlinear stage in the development of modulation instabilities depends on the asymptotic behaviour of the initial disturbance as $|x| \to \infty$. If the perturbation is periodic in space, the sinusoidal modulation waves that build up as a result of a modulation instability will undergo nonlinear distortion: one or more solitons will be formed in the modulation period, but then the solitons will be smoothened and

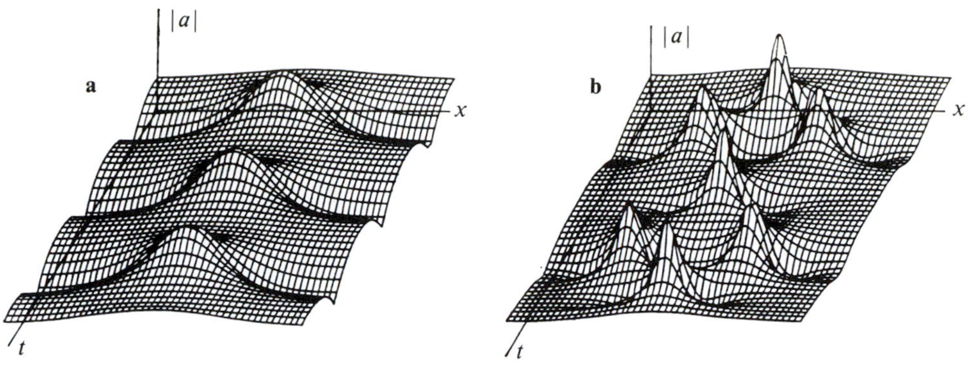

Fig. 2.46a,b. Modulation waves on the surface of a deep fluid [2.75]: *a*) stationary waves; *b*) recurrence

the wave will acquire its initial state, the whole process will then be repeated, and so on. This is precisely the way in which modulation waves on the surface of a deep fluid behave, see Fig. 2.46. A remarkable and astonishing phenomenon! It would, indeed, surprise us if we had no "nonlinear experience" and had never observed something very similar: just recall Fig. 2.12 which illustrates the behaviour of a periodic disturbance in a nonlinear chain or a one-dimensional "medium". Exactly the same thing happens – the sinewave turns into a periodic sequence of solitons, a cnoidal wave, which then evolves back into a sinewave, and so on, i.e. a recurrence effect is observed. The physical explanation of this similarity is quite easy. The nature of the nonlinear evolution is determined both for simple waves "without filling" (waves of the field itself) and for modulation waves by two competing effects – nonlinear contraction and dispersion spreading. The shape and other parameters of the periodic modulation wave must be adjusted so that these effects equalize one another exactly throughout the entire space. Such solitary modulation waves do exist: they are stationary modulation waves that were first investigated in 1966 [2.74]. This is, however, an exception. For all other periodic disturbances, the "contraction" and "spreading" effects predominate alternatively, in the same way as kinetic energy becomes potential energy and vice versa in the oscillations of a pendulum. This is what determines the periodic evolution of a disturbance with periodic boundary conditions that develops as a result of modulation instability.

Formally, the mathematical aspect of the recurrence effect of modulation waves in self-focussing media (or in media with modulation instability) follows from the full integrability of the nonlinear Schrödinger equation with periodic boundary conditions [2.71]. In this case the nonlinear wave has a discrete spectrum (because of dispersion higher harmonics can be considered as nonresonant ones and, consequently, the spectrum as limited), and a mode description can be used for better understanding of the recurrence mechanism. In the simplest case, modulation instability results in resonant interaction of only three modes: the carrier and satellites symmetric about it. The equations for their intensities A_0 and $A_1 = A_2$ bear a close resemblance to the truncated equations of a spring pendulum (cf. (2.5)):

$$\begin{aligned}
&\dot{A}_0 = 2A_0A_1 \sin\Phi \; ; \quad \dot{A}_1 = -A_0A_1 \sin\Phi \; ; \\
&\dot{\Phi} = s + A_1 - A_0 + (2A_1 - A_0)\cos\Phi \; ; \qquad (2.38) \\
&\Phi = [\Delta\omega t + 2\arg(a_0/a_1)]\mathrm{sgn}\alpha \; ; \quad S = \mathrm{sgn}(\Delta\omega\alpha) \; .
\end{aligned}$$

Using the variables $X = \sqrt{2A_0}\cos(\Phi/2)$ and $Y = \sqrt{2A_0}\sin(\Phi/2)$ the phase portraits of the partially integrated system (2.38) are presented in Fig. 2.47: almost all motions are periodic, which is consistent with periodic energy exchange between the satellites and the carrier.

In a less trivial case, when many satellites grow simultaneously as a result of modulation instability, everything is essentially similar except that the shape of the nonlinear wave may be quite complex at the intermediate stage [2.75].

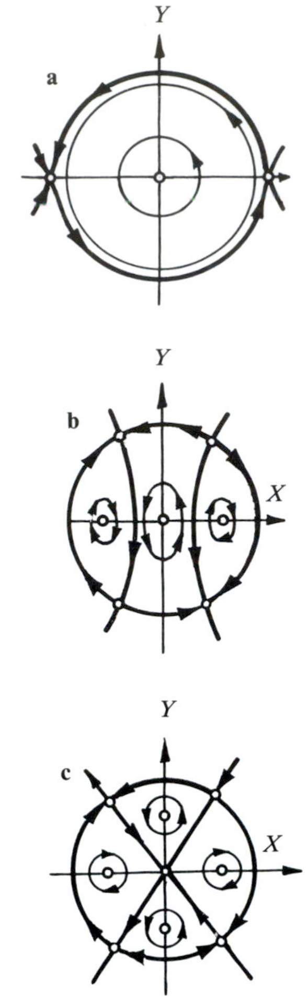

Fig. 2.47a–c. Phase portraits of the system (2.38) with $s = 1$ and $E = A_0^2 + 2A_1^2$:
a) $E < 1/2$;
b) $1/2 < E < 2$;
c) $2 < E$

2.5.6 Modulation Solitons

The analogy that we have established in the behaviour of periodic field waves and modulation waves in nonlinear media can be extended to nonperiodic waves and, in particular, to solitons. We shall soon see that the solitons of the nonlinear Schrödinger equation (in the context of electrical circuits they are in translations from Russian into English often referred to as radio solitons) behave like video solitons, specifically, KdV solitons.[28] Experiments with radio solitons (corresponding to classical ones on deep water) indicate that their parameters do not change in collisions with each other or when they overtake each other; only the phase (of the filling) is altered. Radio solitons prove to be stable formations within the framework of one-dimensional theory.

However, most modulation solitons, like field solitons (see [2.76]) are unstable with respect to multi-dimensional perturbations.

In particular, a waveguide channel of infinite length and a stationary packet with infinite front dimensions that has a finite wavelength in the propagation direction are unstable. This can be made transparent by illustrative though not quite rigorous energy considerations [2.70]. Let $\varepsilon_H(|a|^2)a = \alpha|a|^2 a$ hold in (2.31); then the field under consideration is characterized by the energy

$$\begin{aligned}
H = \int d\boldsymbol{r} \Bigg[& \frac{iv}{2}(a^*\nabla a - a\nabla a^*) \\
& + \frac{1}{2}\frac{d^2\omega}{dk^2}(\partial_x a)^2 + \frac{v}{2k}|\nabla_\perp a|^2 + \alpha|a|^4 \Bigg] \; . \qquad (2.39)
\end{aligned}$$

Besides, (2.33) has another integral, $N = \int |a|^2 d\boldsymbol{r}$, that yields the number of quasi-particles (quanta) in the wave. Let the wavepacket have the length 1 and the number of particles $N = \int |a|^2 d\boldsymbol{r} \approx a^2 l^m$, where m is the dimension of the packet. Then, taking into account

[28] Nearly all effects known in the theory of nonlinear waves without "filling" or internal structure are also observed for modulation waves. Besides, mention must also be made of simple and shock modulation waves.

that the number of particles in the packet is conserved, we have for its amplitude $a(t) \approx \sqrt{N} l(t)^{-m/2}$ and for the energy

$$H \approx \left(d^2\omega/dk^2\right)\left(N/l^2\right) - \left(\alpha N^2/l^m\right) \ .$$

Here the first term is responsible for diffraction spreading of the packet and the second term for its nonlinear contraction. This expression shows that there exists a scale $l_0 = (d^2\omega/dk^2)2/\alpha N$ for which the packet energy will be minimal ($\partial_l H|_{l_0} = 0$) in the one-dimensional case and we may hope that a soliton with these parameters will be stable. The behaviour of the two-dimensional pulse $m = 2$ depends, obviously, on the initial conditions: if $\omega''_{kk} > \alpha N \approx \alpha|a|^2/l^2$, then the energy minimum is reached as $l \to \infty$ and the pulse spreads; but if $\omega''_{kk} < \alpha|a|^2/l$, the energy is minimal as $l \to 0$ and the soliton contracts to a point or collapses. The evolution of a three-dimensional soliton should also end in a collapse: when $m = 3$, nonlinear contraction predominates over diffraction spreading.

Bearing in mind the stability of radio solitons in one-dimensional systems it is natural to use such solitons as an undisturbed (unmodulated) solution for the investigation of a broad range of models similar to the "standard" model – the nonlinear Schrödinger equation. Let us analyse the behaviour of one of these models, namely,

$$\partial_t a - \frac{i}{2}\partial_{xx} a - |a|^2 a = \gamma(a + \partial_{xx} a) - \varrho|a|^2 a \qquad (2.40)$$

which describes the nonlinear evolution of modulated waves in nonequilibrium media (the term proportional to γ denotes a spectrally narrow increment of the waves and $\varrho|a|^2 a$ their nonlinear damping). In the approximation of small damping and a spectrally narrow increment, (2.40) has been derived for Tollmien-Schlichting waves in boundary layers, for Langmuir waves excited by an electron beam, for concentration waves in chemical reactions, etc. (see, for example, [2.77]).

Like the standard conservative model, (2.40) has a solution in the form of an unmodulated harmonic wave with the envelope $a = \sqrt{\gamma/\varrho}\exp[i\gamma t/\varrho]$. When $\gamma\varrho < 1$, this wave is unstable to periodic disturbances with the wave numbers $k < \sqrt{2\gamma(1-\gamma\varrho)/\varrho(1+\gamma^2)}$. At very small amplification and damping ($\gamma, \varrho \ll 1$) the development of this instability results, as in the undisturbed model (for initial disturbances rapidly decreasing at infinity), in a steady-state solution in the form of a sequence of solitons

Mathematical solitons are defined as analytical solutions of (partial differential) equations that collide with each other recovering after the scattering event their asymptotic shapes, amplitudes, speeds – the only trace of the interaction being a possible phase-shift.

Physical solitons are less well defined, often a "lump" is happily referred to as a soliton.

$$a(x,t) = \frac{A\exp\left[it(A^2 + v^2/2)/2 + v(x - vt)/2\right]}{\cosh\left[A(x - vt)/\sqrt{2}\right]} \qquad (2.41)$$

whose amplitude and velocity will now vary in time. Let us find equations for the parameters of such "modulated" radio solitons. For this purpose, we shall determine the rate of change of the particle number $N = \int |a|^2 dx$ and the quasi-momentum $p = \int i(a^*\partial_x a - a\partial_x a^*)dx$, which are the integrals of the conservative model. Then we shall use (2.40) to find equations for $A(t)$ and $V(t)$:

$$\begin{aligned} \frac{dA}{dt} &= 2\gamma A\left(1-\frac{A^2}{6}-\frac{v^2}{4}\right)-4\varrho\frac{A^3}{3} \\ \frac{dv}{dt} &= -2\gamma A^2\frac{v}{3}\,. \end{aligned} \tag{2.42}$$

We see that as $t \to \infty$ all solitons stop and their amplitudes tend to the same value $A \to A_0 = \sqrt{6/[1+(4\varrho/\gamma)]}$. Thus, the initial periodic disturbance evolves into a lattice of modulation solitons.

If the amplification and damping are not too small ($\gamma, \varrho \sim 1$), the initial perturbation evolves in a completely different fashion: complex behaviour arises as a result of the development of modulation instability, see Chap. 5 *Turbulence*.

We have already mentioned that the simple or complex behaviour in a dynamic system is related to its "distance" from the nearest fully integrable system. For (2.40) this system is the nonlinear Schrödinger equation with periodic boundary conditions. In particular, for $\gamma \ll \varrho \ll 1$ the distance between (2.33) and (2.40) is small and the behaviour of (2.40) is simple, but as the distance increases ($\gamma, \varrho \sim 1$), numerical simulation [2.77] indicates that stochasticity arises in the system (2.40) with periodic boundary conditions, see Chaps. 3 *Chaos* and 5 *Turbulence*.

3. Chaos

One of the most remarkable recent progresses in science has been a significant advance in the understanding of randomness. By now we know that in principle any random signal can be reproduced by a nonlinear dynamic system. However, for modelling different random or stochastic processes or signals different dynamic systems are needed, that is systems with different numbers of degrees of freedom. White noise poses the hardest problem for dynamic reproduction since it requires an infinite number of degrees of freedom implying that its dimension has to be infinite too. – We will first make a few historical remarks to go then over to a discussion of chaos.

3.1 Historical Remarks

We have attempted more than once to enter into the realm of nonlinear physics, where dynamic systems (without noise or fluctuations, i.e. "nonrandom" ones) behave in an irregular and intricate fashion, with many signatures of "truly random" behaviour. Complicated irregular features typical of various domains of nonlinear physics have been known for a long time. The nontrivial behaviour of the Van der Pol generator under the action of external forces [3.1] was encountered back in the 1940s. In the 1950s irregular oscillations made themselves known in the analysis of a simple "disk" model of the magnetic dynamo [3.2]. A little earlier, studying a continuous system of temperature control [3.3] *A.S. Alekseev*, one of Andronov's students, observed experimentally period doubling of oscillations and the birth of regimes with complex dynamics. Those were first unintentional contacts with a remarkable new phenomenon that later revolutionized our understanding of randomness – the first encounter with dynamic chaos.

Although quite a few mathematical tools for the description of the nontrivial behaviour of dynamic systems in phase space (e.g. Poincaré mapping, homoclinic structures, etc.) were known as early as the beginning of this century, it was only in 1963 that the meteorologist *E. Lorenz* [3.4] undertook the first purposeful investigation into the irregular behaviour of a dynamic model of a real physical process.

He studied numerically the chaotic behaviour in a model with the equations

$$\dot{X} = \sigma Y - \sigma X \; , \; \dot{Y} = rX - Y - XZ \; , \; \dot{Z} = -bZ + XY \qquad (3.1)$$

designed to describe thermoconvection (see below). Lorenz's system became one of the basic models of chaotic self-excited oscillations applied in various domains of nonlinear physics. For example, *Haken* and *Graham* [3.5] reduced to (3.1) the equations for the single-mode generation of a solid-state laser. *Rabinovich* [3.6] showed that (3.1) can be interpreted as the set of equations depicting a nonlinear oscillator whose frequency depends inertially on the oscillation energy.

We would like to emphasize that it was only ten years after the publication of Lorenz's paper that (3.1) attracted a wider interest. Evidently, this is explained by the fact that – besides static and periodic or quasi-periodic motions – no similarly sophisticated motions were studied at the beginning of the 1960-s. However, in the fifteen years to follow, owing greatly to the contributions of mathematicians (such as Anosov, Sinai, Smale, Williams, and others), it was shown that stochastic behaviour is persistent in dissipative systems having phase space dimensions greater than two in closed domains of parameter space. This insinuates that stochastic behaviour is neither an exception nor an anomaly. When it was eventually understood that stochastic motions are in deterministic systems as abundant as periodic ones, models of "chaos" became ubiquitous.

In the last few decades, "stochastic dynamics" became increasingly popular among physicists. This is due to both the awareness of a great number of new specific problems in different branches of physics and the opportunity to advance further the fundamental problem of the relation between dynamic and static laws of physics which were traditionally considered to be opposed to each other. The stochastic motion of simple dynamic systems is responsible, for example, for the loss of charged particles in accelerators and plasma traps [3.7] and for similar phenomena in the Earth's radiation belt [3.8]; for stochastic heating [3.9] and acceleration of particles by a periodic field [3.10] which are now rather well studied so that they have even found practical applications; it is possible to construct a solid-state intrinsic noise generator with spectrum restructuring [3.11]. Even the good old celestial mechanics exhibits quite a number of irregular motions, the most unexpected one being the chaotic orbiting of Saturn's satellite Hyperon [3.12].

We would like to add that by now the relation of stochastic self-oscillations to turbulence in fluid mechanics has been proved experimentally, see Chap. 5 *Turbulence*.

3.2 Marble in an Oscillating Chute

Let a marble roll without friction over the bottom of the chute depicted in Fig. 2.36b. For small energies, the marble will oscillate in one of the (relative) minima or valleys, while in case of large energy it will move from one valley to another periodically in time. We are well acquainted with the phase portrait of such a nonlinear oscilla-

... have finally realized that most dynamical systems do not follow simple, regular, and predictable patterns, but run along a seemingly random, yet well-defined, trajectory. The generally accepted name for this phenomenon is chaos ...

M.C. Gutzwiller
"Chaos in Classical and Quantum Mechanics"
(Springer, New York, Berlin, Heidelberg 1990)

Fig. 3.1. Marble in oscillating chute

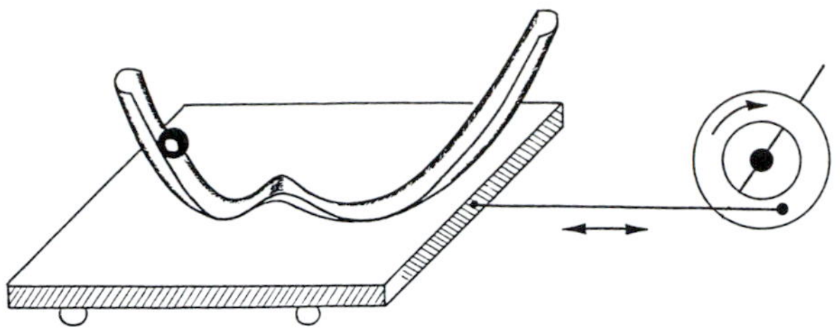

tor, see Fig. 2.36b. – But what happens if the chute itself oscillates according to a harmonic law as illustrated in Fig. 3.1? If we were interested in a linear system with no resonances, the two oscillations would simply add up. However, in our case, the initial oscillation energy will determine the further details of the situation. If the initial energy is small, the behaviour will be very much like that of a linear system: quasi-periodic oscillations or (in the presence of resonances) a building-up of oscillations, with their subsequent suppression when getting off-resonance (nonlinear resonance). The motion may also be quasi-periodic or periodic. For larger initial energies we will encounter a very different behaviour; even for a small amplitude of the chute's oscillations, that is for a small amplitude of the harmonic force on the right-hand side of the system

$$\ddot{x} - \left(x - x^3\right) = f_0 \sin\theta \ , \quad \dot{\theta} = \omega \ , \qquad (3.2)$$

describing the motion of the marble.

... classical chaos is rough and fractal, whereas quantum chaos is smooth and elusive ...

M.C. Gutzwiller
ibid.

System (3.2) has a three-dimensional phase space, with the coordinates $\dot{x}$, x, and θ; θ is periodic in 2π. We can reduce the dimension of the space by using the method of Poincaré maps and going over from the analysis of the trajectories in the initial space to the analysis of their "traces" in the intersecting plane $\dot{x}$, x; recall Fig. 2.7. In this plane periodic or fixed points correspond to periodic motion and closed tori to quasi-periodic motion.

For a small amplitude f_0, oscillations with sufficiently high initial energy (symmetric motions) and with low energy will only be slightly distorted. But motions with the energy U_{cr} equal to the height of the potential barrier between two "valleys" are qualitatively different and the singular trajectory corresponding to them for $f_0 = 0$ – the "eight" in the section plane – will be destroyed. The future of the marble that has climbed up from the left and reached the top of the hump of the valley depends actually on the momentum imparted to it by the external force. If at that very instant the chute swings to the right, the marble will roll back to the bottom of the valley, but if the chute swings to the left, the marble will roll over the hump and slide down into the neighbouring valley. Since the period of the chute's oscillations is in no way connected with the period of the marble's oscillations (in particular, the "period" is infinite in the motion along the "eight"), a marble whose energy is close to U_{cr} will move irregularly from one valley to another.

Mathematically this corresponds to the destruction of the separatrix and the formation of a homoclinic structure in the phase space $\dot{x}$, x, θ. The existence of a homoclinic structure in the presence of f_0 can be checked as follows: When there are no perturbations, the separatrix originating from the saddle equilibrium state coincides with the one entering this state. A perturbation would split them. Figure 2.13 demonstrates what may happen in such a case. We can see that having intersected once, the separatrices will intersect an infinite number of times, a typical feature of homoclinic structures. Therefore, there are stochastic oscillations in the system if perturbational calculations of the distance between the separatrices yield a change of its sign in the motion along the unperturbed separatrix (Melnikov's criterion) [3.13].

Let us see what happens if we bring a "drop of a phase liquid" into the neighbourhood of a destroyed separatrix. When moving along the trajectories the drop will be deformed in a complicated manner: it contracts in some directions and stretches into others.[1] For $t \to \infty$, the drop will spread out to form a thin stochastic layer near the separatrix, see Fig. 3.2. Outside the stochastic layer(s), the motion will continue to be regular. This is an example of a system possessing a separated phase space. The region of stochasticity is separated from the region of nondestroyed tori. A similar picture is observed in the four-dimensional phase space of two coupled oscillators (remember the example given by Hénon and Heiles, see Figs. 2.6, 7). – In this case the integral energy surface will be a three-dimensional one.

There are no geometrical (topological) rules ensuring the separation of chaotic and regular motions in the phase space of a Hamiltonian system with the dimension greater than four. The regimes of stochasticity in different regions of phase space may be interconnected

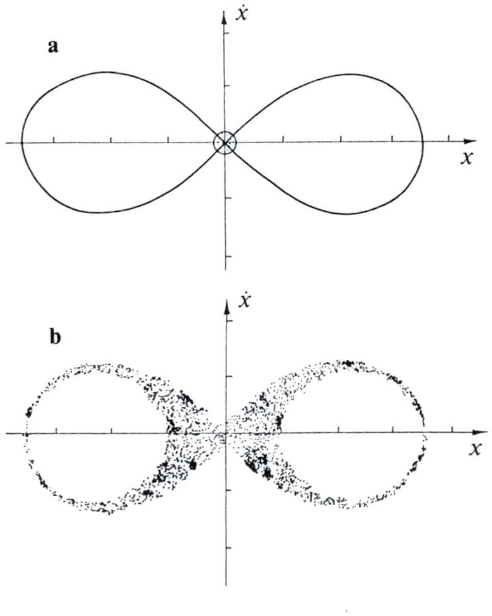

Fig. 3.2a,b. Spreading of drops of a phase fluid on a homoclinic structure (Poincaré section); the undisturbed separatrix is denoted by a full line: *a)* $t = 0$; *b)* $t \to \infty$

[1] We recall that we are here concerned with a Hamiltonian system implying that the initial phase volume will be conserved.

by "pieces" of the same trajectory. Usually this is accomplished by trajectories close to the separatrices (stochastic diffusion or *Arnold's* diffusion [3.14]).

As the amplitude of the external force grows, the region (or regions) of stochasticity in the phase space of a Hamiltonian system will expand. This corresponds to the expansion of the regions of nonlinear resonances at different combination frequencies. When these regions of nonlinear resonances have overlapped, the stochastic motion will occupy the dominant part of phase space (*Chirikov's* criterion of resonance overlapping [3.15]).

The motion corresponding to an imbroglio behaviour of the trajectory within a bounded phase volume has all most characteristic features of a random process. Let us explain this point qualitatively. Complex deformation of the initial phase volume – the "drop of the phase liquid" and its mixing in the neighbourhood of the separatrix – are the consequences of a local instability of nearly all trajectories inside the homoclinic structure. Points, that were at the moment $t = 0$ arbitrarily close to each other diverge with increasing t. Such an instability of trajectories enclosed in a bounded volume yields a complicated, imbroglio motion, making it indistinguishable from a "truly" random one. Indeed, because of the boundedness of the phase volume, nearly any open trajectory will approach itself arbitrarily close in a sufficiently large period of time. But in view of the instability, this proximity implies by no means that the following stage will be similar to the previous one. On the contrary, a small perturbation will grow and the subsequent route of the system's motion will be almost independent of its past. Just as in the case of the "truly" random walk of a Brownian particle.

Instability: small causes may give rise to dramatic changes.

These considerations lead us to another important manifestation of instability: it is impossible to reproduce the motion of a system with unstable trajectories by specifying its location with an arbitrarily high but finite precision at a certain moment of time [3.16, 17].

Thus, the occurrence of randomness in the behaviour of a "nonrandom" nonlinear system is related to two facts: first, in a definite sense, nearly all open trajectories inside a bounded phase volume are random; second, in a natural way there emerges the concept of an ensemble which is well-known from applications in probability theory. In our case, it is an ensemble of various portions of unstable trajectories inside a bounded phase volume.

We should add that the similarity of the trajectory of a dynamic system within a bounded phase volume to the one of a Brownian particle is not a mere analogy. There exist dynamic systems [3.18] for which the probability distribution at any sufficiently great period of time approaches the probability distribution of a Brownian particle, that is a Gaussian distribution (which is known to be the manifestation of the law of great numbers in probability theory).

3.3 Stochastic Self-Excited Oscillations

We shall now consider the problem of the marble in a chute taking into account dissipation. At first sight we would expect a simple motion, i.e. no chaos. What we observe is that the phase volume of the system

$$\ddot{x} + \gamma\dot{x} - \left(x - x^3\right) = f_0 \sin\theta\ , \quad \dot{\theta} = \omega\ , \tag{3.3}$$

is no longer conserved. It is compressed and, consequently, the trajectories mixing in a three-dimensional phase volume would be expected to stay close to the surface where they cannot mix. However, these considerations are not true. From the compression of the phase volume it only follows that as $t \to \infty$ the dimension of the limit set (attractor) must be lower than three; but it does by no means follow that it must equal two (in the latter case chaos would indeed be impossible). The dimension of the attractor in a dissipative system is always smaller than the dimension of its phase space. However, its dimension D may also be a fractal one, in particular, it may be $2 < D < 3$. For $D > 2$ stochasticity of the motion is retained.

The dimensional characteristics of chaotic motion are very important and highly informative, therefore we shall investigate them in more detail later on. But before doing so, we shall consider a few examples of the stochastic behaviour of dissipative systems and discuss the relation between "Hamiltonian" and "dissipative" chaos.

The periodic trajectory of a Hamiltonian system may go over into a limit cycle (saddle). In the presence of arbitrarily small dissipation and an energy supply compensating for the dissipative losses the stochastic set of a Hamiltonian system may in a similar situation neither break up nor disappear, but simply become an attracting set, that is transform into a stochastic attractor. Strictly speaking, this does not occur in reality: islands of regular behaviour inside the homoclinic structure contain periodic trajectories, which go over into stable cycles when dissipation is introduced (this holds, at least for systems with one and a half or two degrees of freedom). For $t \to \infty$ nearly all trajectories in the neighbourhood of the homoclinic structure will tend to these stable cycles. In other words, when the conservative character of the system is slightly disturbed (by introducing dissipation plus energy supply), the chaotic dynamics corresponding to the walking of the image point on the homoclinic structure must become transient, that is unobservable for $t \to \infty$.

Nevertheless, our intuitive concepts are to a large extent correct. Actually, with a further increase of dissipation, the stable cycles may merge with the unstable ones (which, in the neighbourhood of the homoclinic curve, form a countable set) and disappear. Then an attractive stochastic set containing exclusively unstable trajectories will emerge in the phase space of the dynamic system. Ruelle and Takens coined the name strange attractor for it.

... the universe is not only queerer than we suppose but it is queerer than we can suppose.

J.B.S. Haldane
"Possible Worlds and Other Papers"

(Chatto and Windus, London 1945)

The possibility of the existence of attracting sets containing only unstable trajectories seems to be paradox. Indeed, a consequence would then be that unstable motions should be observed at large times, includ-

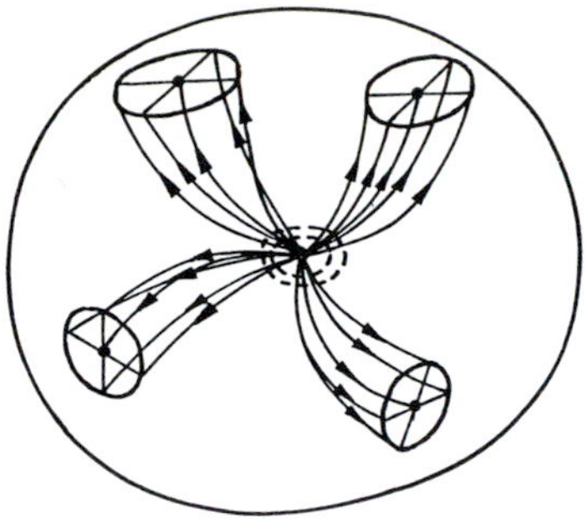

Fig. 3.3. Unlimited growth of a phase volume containing only trajectories that are unstable into all directions

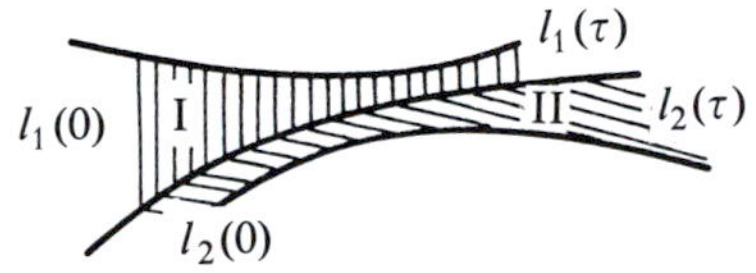

Fig. 3.4. Example of a saddle trajectory; direction *I* is stable and direction *II* unstable

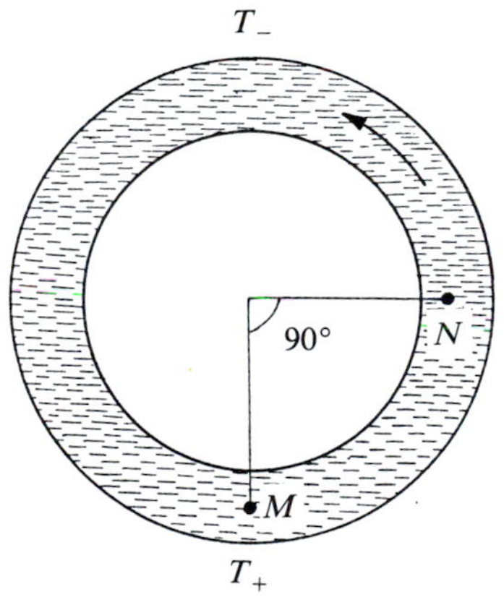

Fig. 3.5. Thermoconvection in a toroidal cavity.
A toroidal cavity is filled with an incompressible liquid whose density ϱ depends on the temperature T as $\varrho = \varrho_0[1 - \alpha(T - T_0)]$ with $\alpha > 0$. The cavity is placed vertically in a homogeneous field of gravity g. The angular coordinate φ is measured counter-clock-wise with regard to g. The temperature at the cavity walls does not change with time and does not depend on φ; at the bottom of the cavity the temperature is larger than at the top (it is measured at the points M and N)

ing $t \to \infty$. At the same time, assertions such as "unstable trajectories are of no physical interest because even very small disturbances would destroy them" would at first sight appear to be well justified. The solution of this paradox is as follows. The system moves for a finite time in the vicinity of the unstable trajectory and goes then away into the neighbourhood of another trajectory. However, these transitions are accomplished in such a way that at any moment of time the system is moving in the neighbourhood of an unstable trajectory (continuously refined in the process of motion) – the theorem of ε-trajectories [3.19].

In this case small perturbations may affect the local route of the system, but they are unable to change the global characteristics of the motion for large times which are determined by the very existence of the attractor and by its statistical properties. We have many confirmations for the realization of such a behaviour of dynamic systems: in long-term weather forcasts, in ecology, and, perhaps, even in history.

In many models of real physical systems one observes "nonrigorous" strange attractors, that is attractors containing stable periodic trajectories whose basins of attraction are so small that neither computer nor physical experiments are able to detect them. It should be noted that homoclinic structures are inherent to any strange attractor.

The mathematical compatibility of the instability of all the trajectories belonging to the stochastic set with the fact that this set is an attractor, attracting all trajectories found in its neighbourhood but not being part of it, has the following explanation: The trajectories forming the strange attractor cannot be unstable in all directions simultaneously since this would lead to an unbounded phase volume, in which the unstable trajectories are located, Fig. 3.3. However, the attractor is localized in phase space, i.e. its unstable trajectories are located in a bounded region. This is only possible when the trajectories are unstable in some directions but stable (attracting) in others – saddle trajectories, see Fig. 3.4. If we now suppose that the stable directions of all saddle trajectories are oriented "outwards" the attractor, then we can imagine, although rather crudely, an attracting structure of unstable trajectories – a strange attractor.

3.3.1 The Lorenz Attractor

In view of the universality and wide applicability of the famous Lorenz system (3.1) we elaborate on the attractors in phase space focussing our attention on the strange attractor known as the *Lorenz attractor*.

For the sake of clarity, we shall discuss a specific physical situation described by (3.1). As an example we consider self-excited oscillations in a ring resonator, a vertically arranged closed tube in a homogeneous gravity field, filled with a viscous fluid [3.20], see Fig. 3.5. When it is heated from below, a convection regime is established: the fluid heated at the base of the ring will be lighter and thus rises, while the cooler fluid portion from the top moves downwards. Thus, starting with some temperature difference $T_- - T_+ = T_{\text{top}} - T_{\text{bottom}} = \Delta T_1$, a regime of stationary clockwise or counter-clockwise rotation is established in the fluid. The entire liquid is now rotating as a whole – a single, large-scale motion is observed. When $\Delta T_2 > \Delta T_1$ an increase in ΔT will result in self-excited oscillations: the liquid ring in the tube will change its direction of rotation every now and then. Physically, this phenomenon may be explained as follows. Let the liquid move clockwise at a given moment of time, for sufficiently large ΔT Archimede's force acting on it is rather large and the water ring is so much accelerated that the cooler fluid elements from the top traverse so quickly through the heated base that there is not enough time for them to get heated up, thus they can no longer climb up to the top of the ring and the motion comes to a halt (since Archimede's force alone is insufficient to surmount the resistance of viscosity and gravity). Now the lower (right) portion of the fluid is warmer and, consequently, lighter than the one at the top. As a result of the slowing down of the ring's motion, the fluid at the base of the ring is heated up and rises – but now in the opposite direction, since the pressure on the right is smaller than on the left. Thus, the fluid ring changes its direction of rotation and moves now counter-clockwise. Subsequently everything repeats in reverse order. These self-excited oscillations due to thermal convection may be periodic or stochastic. They do not involve other scales of motion, besides the basic one. Therefore, the mathematical model describing them can be obtained from the basic equations of hydrodynamics, see Fig. 3.5.

We obtain for the dimensionless velocity of the liquid ring $X(t)$, the system (3.1) of ordinary differential equations. The fluid temperature at point N is denoted by $Y(t)$ and the one at M by $Z(t)$, see Fig. 3.5. The parameters σ and r in (3.1) are positive.

The bifurcation diagram for the changes of the regimes of the Lorenz system is depicted in Fig. 3.6. We shall briefly describe these bifurcations. They have been investigated using a computer analysis of the Poincaré mapping on the section plane Σ transversally oriented to the Z-axis through the equilibrium states $C_{1,2}$, see Fig. 3.7. This two-dimensional mapping appears to be strongly compressed into one direction (Θ on Σ; see Fig. 3.8) and stretched into another. Repeated applications of this mapping, transform each cell on Σ to "lines" having a thin Cantor structure. Therefore we can restrict our analysis to a one-dimensional mapping of the lines into themselves and onto one another. Figure 3.9 shows the behaviour of the unstable separatrices of the zero equilibrium state (saddle), that determines the properties of such a one-dimensional mapping [3.21].

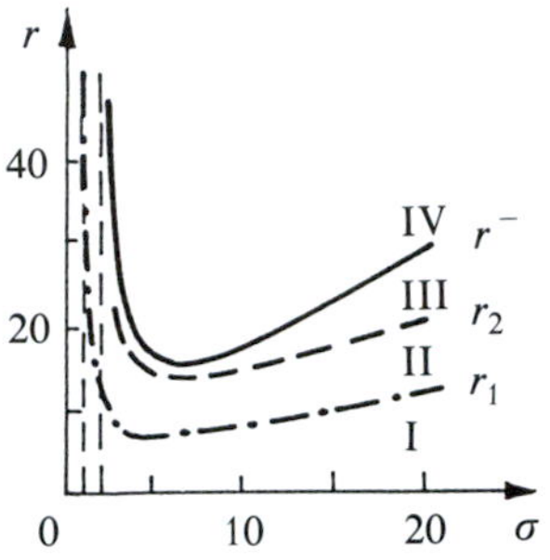

Fig. 3.6. The bifurcation diagram of a Lorenz system:
I – without convection;
II – stable convection (for different initial conditions the liquid rotates into different directions);
III – three stable regimes (for small perturbations stable convections with a rotation and for large perturbations chaotic motion);
IV – only one nontrivial attractor, a strange attractor

For $r < r_1$, the separatrices describe damped pulsations in which the initial oscillation phases are retained. At $r = r_1$, the unstable separatrices touch the stable two-dimensional separatrix AB, see Fig. 3.8. For $r > r_1$, they go already over from "their" stable focus C_1 or C_2 to a "strange" one. Simultaneously, the separatrix loops give birth to two symmetrically located unstable limit cycles. For $r > r_2$, the separatrices tend already to these newly born cycles, rather than to the equilibrium states C_1 and C_2 which are still stable: this is the instant of birth of yet another attractor (besides C_1 and C_2) – a strange attractor.

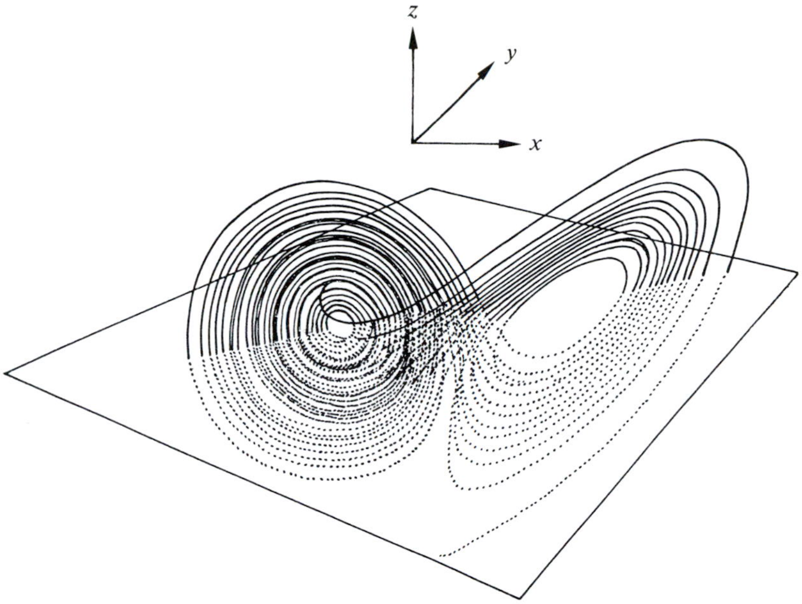

Fig. 3.7. Lorenz attractor

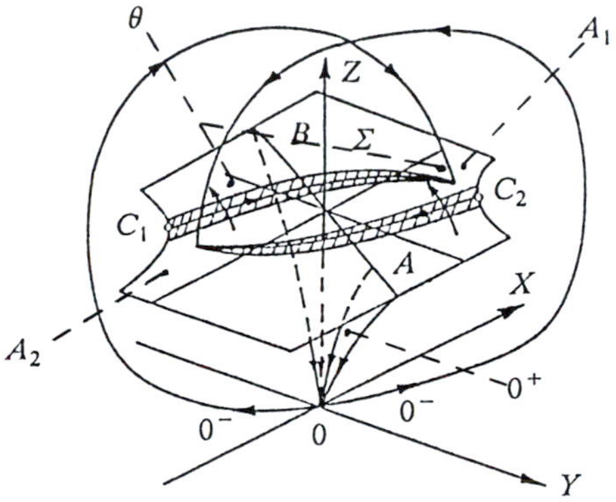

Fig. 3.8. Phase space of a Lorenz system

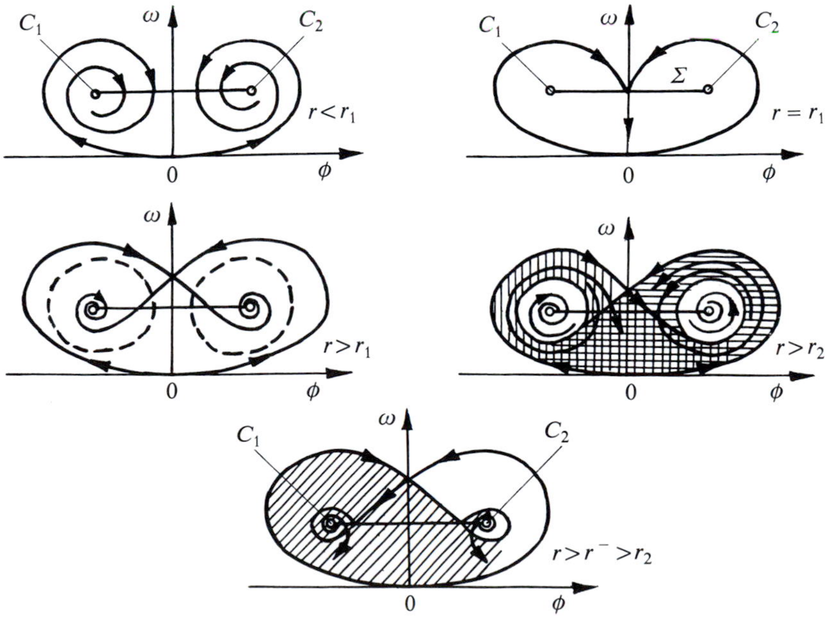

Fig. 3.9. Behaviour of an unstable separatrix in its projection onto the plane

Inside this attractor bounded by unstable separatrices and cycles the trajectories behave in a very complicated manner, see Fig. 3.7 depicting a Lorenz attractor (corresponding to the oscillogram $\nu(t)$ shown in Fig. 3.10). This complexity is related to the fact that a countable set of unstable cycles, or saddle cycles belongs to this attractor (responsible for its emergence is the homoclinic structure that has appeared prior to the strange attractor). On a stable separatrix the trajectory approaches one of the cycles, makes several turns in its vicinity, then leaves it, heads for another one, spins in its vicinity and so on.

A physicist refers to a pure tone as a single sine-like oscillation where the frequency spectrum consists of a single line.

Besides the strange attractor there are for $r \gtrsim r_2$ two more "non-strange" attractors. Therefore, it depends on the initial conditions whether a static regime or one with stochastic pulsations will be established. In other words, in this parameter region we have a hard onset of chaos. As $r > r_2$ increases, the radius of the unstable cycles is reduced and for $r = r_1$ they are "trapped" into the equilibrium states C_1 and C_2 transferring their instability to them, i.e. in the phase space (3.1) there remains only one attractor – a strange attractor.

3.3.2 Synchronization – Beats – Chaos

So far we only know the behaviour of the Van der Pol generator under the action of a harmonic force in the case of small nonlinearity in system (2.22) and weak external forces. We have observed two effects: synchronization as frequency trapping by the external force and stable quasi-periodic oscillations (pulsations) when going beyond the synchronization region. However, the phenomena observed in a nonautonomous Van der Pol system are much more diverse when these parameters are no longer small. More than forty years ago *Cartwright* and *Littlewood* [3.22] found that for large f and μ the phase space of the system $\ddot{x} - \mu(1 - x^2)\dot{x} + x = f_0 \sin \omega t$ contains two stable periodic trajectories and a homoclinic structure with a countable number of unstable periodic trajectories. Although the attractors in this system correspond to periodic oscillations, the transition regime, during which the image point walks within the homoclinic structure, is very complicated and long, see Fig. 3.10.

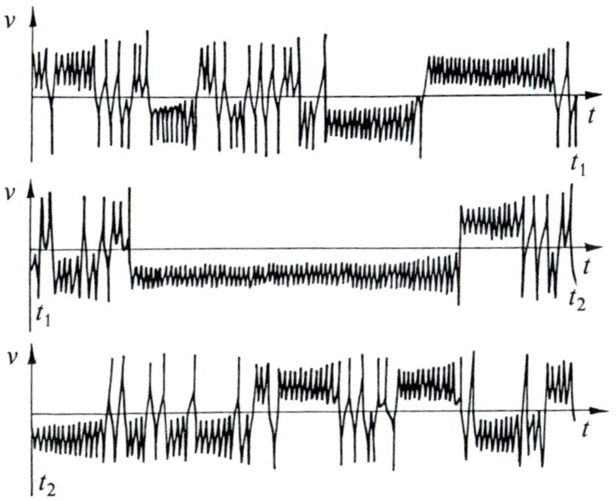

Fig. 3.10. Oscillogram of oscillations in a Lorenz system

This nonattracting chaos becomes an attracting one if we can dispose of the stable cycle(s). Due to resonance overlapping this is accomplished in the forced Van der Pol – Duffing system

$$\ddot{x} - \left(1 - x^2\right)\dot{x} + x^3 = f_0 \sin \omega t \ . \tag{3.4}$$

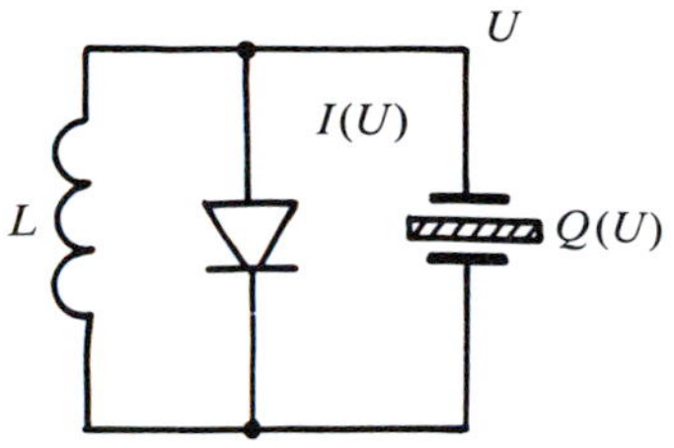

Fig. 3.11. Generator with nonlinear capacity (modelled by the Van der Pol – Duffing equation)

This system describes a generator with a nonlinear capacitor or inductivity, see Fig. 3.11. In the phase space of (3.4) there exists for a wide range of parameters a strange attractor (see Fig. 57 in [3.23]).

It is remarkable that *Van der Pol* himself and his co-author *Van der Mark* [3.1] were the first to observe and describe the irregular behaviour of a nonautonomous generator. They experimented on a neon-lamp nonautonomous generator designed to generate subharmonics. In order to observe the oscillation frequencies, the researchers simply listened to the noise by ear-phones switched to the circuit (this was back in 1927). With increasing detuning between the frequency of natural self-oscillations and the external force, they observed consecutively synchronization regimes at the subharmonics 1/2, 1/3, 1/4, etc. However, inbetween the synchronization bands they heard at different subharmonics "irregular noise"! These experiments were repeated in 1986 [3.24] so that we now can say for sure that Van der Pol and Van der Mark heard dynamic chaos.

Under the action of a periodic force nearly any nonlinear oscillator with dissipation may become a stochastic generator. If there are several stable equilibrium states on the phase plane of an oscillator (i.e. if the potential of the oscillator has several potential wells), then relatively small external forces will give rise to chaotic self-excited oscillations in the form of randomly mixed sequences of oscillations in the different wells, see (3.2). An ordinary pendulum with friction and a periodic force may also turn into a stochastic or noise generator. The strange attractor of such a pendulum is shown in Fig. 3.12 [3.25].

A sound contains several eigenmodes and a more sophisticated frequency spectrum with several lines.

The four characteristic properties of a sound are its pitch, loudness, duration and timbre (quality or tone colour).

Similar random walks are also observed in oscillators with non-parabolic "single-well" potentials. However, in this case the role of equilibrium states is played by the newly born periodic saddle trajectories of the nonautonomous system. It is this regime of stochastic self-excited oscillation that is observed in the nonautonomous anharmonic dissipative oscillator

$$\ddot{x} + \gamma\dot{x} + x + x^3 = f_0 \sin \omega t \ , \tag{3.5}$$

whose strange attractor, together with the trajectories attracted by it, are displayed in Plate 7 of Chap. 6 *Nonlinear Physics* [3.23]. In such "mono-stable" oscillators in the self-oscillation regime chaos arises usually at significantly larger amplitudes of the external force compared to "multi-stable" oscillators.

3.3.3 Autonomous Noise Generator

We are now able to synthesize such a generator. Let us consider the *Van der Pol* (VdP) generator. We know that there are at least two pre-

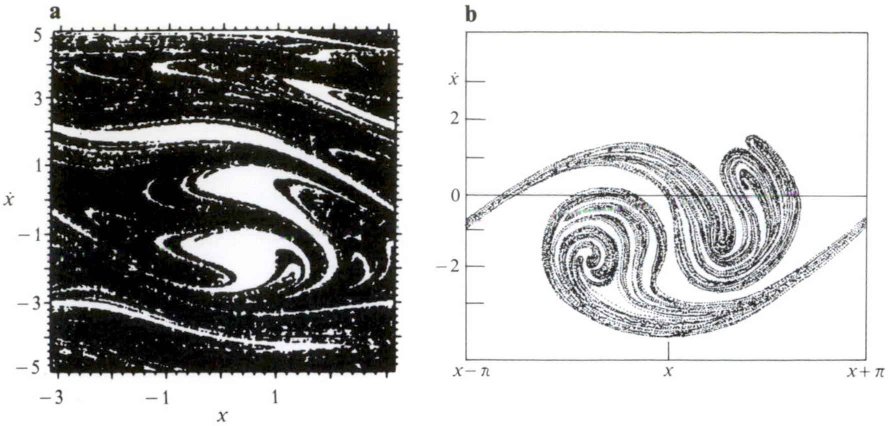

Fig. 3.12a,b. Phase portraits on the secant of the damped driven pendulum with $\ddot{x} + \gamma\dot{x} + \omega^2 \sin(x) = f\cos(\Omega t)$:
a) no strange attractor; complex structure of the basin of attraction, stable periodic regime ($\omega = \Omega = 1$; $\gamma = 0.1$; $f = 1.2$);
b) strange attractor ($\omega = 1$; $\Omega = 2/3$; $\gamma = 0.5$; $f = 1.4$)

conditions for the onset of chaos: the instability of individual motions and the boundedness of the region of phase space (with a degree of freedom larger than two) containing these motions. At small amplitudes, the oscillations in the Van der Pol generator are unstable (in the phase plane these oscillations correspond to the trajectories diverging from the equilibrium state). If the trajectories corresponding to the growing oscillations are "taken out" of the space and then "returned" to the neighbourhood of the coordinate origin, then we will observe a trajectory mixing in the bounded region considered.

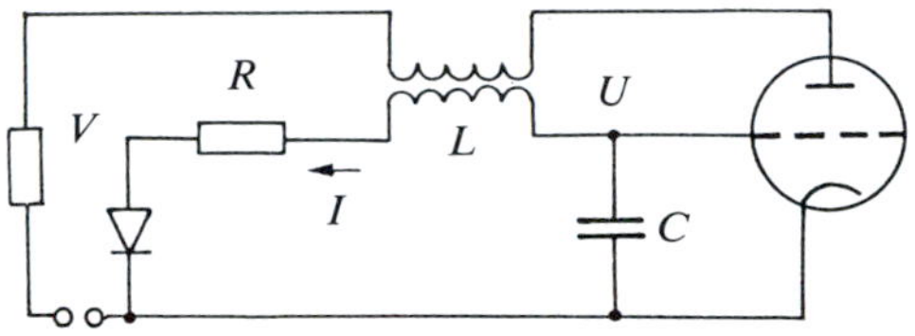

Fig. 3.13. Block scheme of a simple noise generator

The phase space of a self-excited oscillator may be arranged in such a manner by adding to the traditional Van der Pol circuit, for instance, (like in Fig. 3.13) a tunnel diode with the Volt-Ampere-characteristic depicted in Fig. 3.14. Neglecting the nonlinearity of the triode, we can write the equation for the oscillator in the form [3.11, 25]

$$\begin{aligned} &\ddot{u}_1 - 2h\dot{u}_1 + u_1 = -\alpha u_2 \ , \\ &\mu\dot{u}_2 = \dot{u}_1 + f(u_2) \ , \end{aligned} \tag{3.6}$$

where $u_1 = I/I_m$, $u_2 = V/V_m$, $t = t_{\text{dim}}/\sqrt{LC}$ and $h = (MS - RC)/2\sqrt{LC}$ is the increment of oscillations in the circuit, $\alpha = V_m/\sqrt{L/C}I_m$ describes the influence of the nonlinear element on the oscillations and $\mu = [(V_m/I_m\sqrt{LC})C_1] \ll 1$ is a small parameter taking into account the spurious capacitance of the tunnel diode.

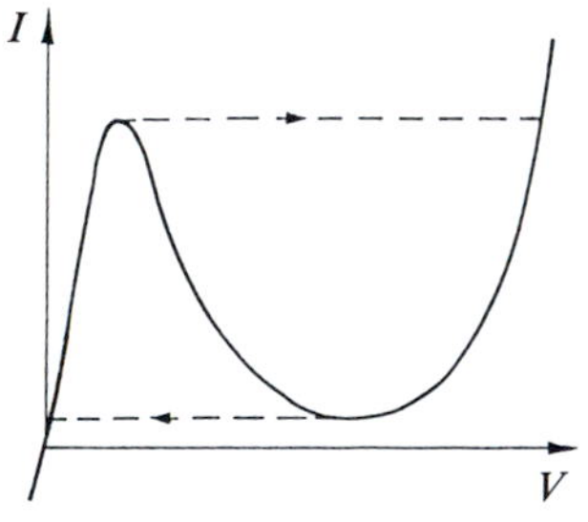

Fig. 3.14. Characteristic of a tunnel diode; $I = I_{\text{td}}(V)$

Qualitatively the operation of this oscillator is as follows. As long as current I and voltage V are small, the tunnel diode does not affect significantly the oscillations that are building-up in the circuit due to the negative resistance of the tube. The voltage in the tunnel

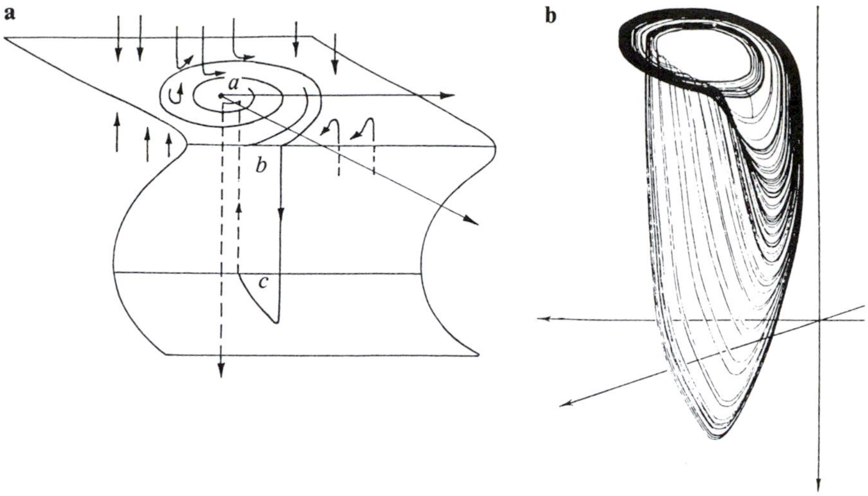

Fig. 3.15a,b. Strange attractor of the system (3.6):
a) structure of phase space;
b) strange attractor of a physical generator

diode $V_{td} = V(I)$ is determined by the left branch of the diode characteristic. The unwinding spiral depicted in Fig. 3.15 corresponds to this stage. When the current reaches the value I_m, the diode switches to establish the voltage V_m. Then, the current I decreases (with the voltage being determined by the right-hand branch of the characteristic) and the diode switches back, see Fig. 3.14. In other words, when the oscillation amplitude in the circuit becomes sufficiently large, the losses increase in a jump and the oscillation amplitude drops. This is illustrated by the section ab of the trajectories in Fig. 3.15. Thus, the generated signal is a sequence of trains of building-up oscillations. This is nicely confirmed by experiment, see Fig. 3.16. Of course, these

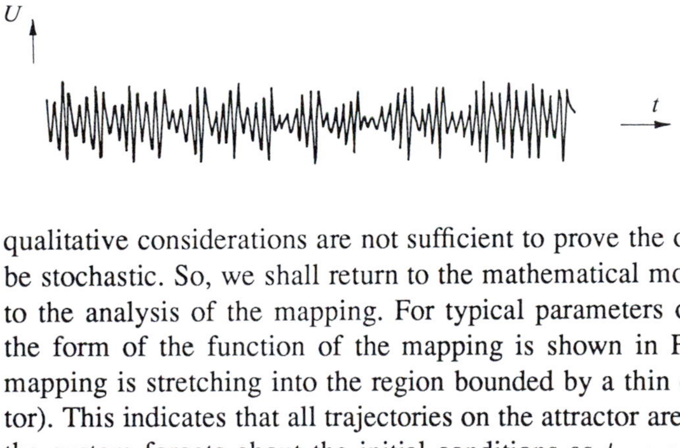

Fig. 3.16. Oscillogram of the oscillations in the system shown in Fig. 3.13

qualitative considerations are not sufficient to prove the oscillations to be stochastic. So, we shall return to the mathematical model (3.6) and to the analysis of the mapping. For typical parameters of oscillators, the form of the function of the mapping is shown in Fig. 3.17. The mapping is stretching into the region bounded by a thin curve (attractor). This indicates that all trajectories on the attractor are unstable and the system forgets about the initial conditions as $t \to \infty$. The possibility of the probability density to assume after multiple application of the mapping some value u tends to an invariant distribution that does not depend on the probability density distribution of the initial fluctuations. We should add, that the statistic characteristics of the regime of stationary generation, which are determined by the dynamics of (3.6), are stable not only with regard to initial perturbations, but also with regard to continuous external fluctuations [3.26]. A noise generator, indeed!

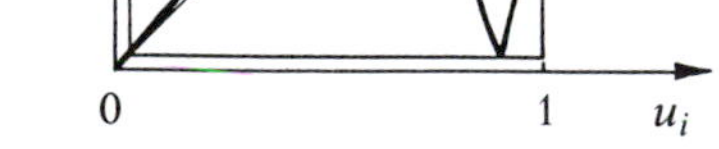

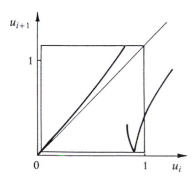

Fig. 3.17. Poincaré map of the system (3.6) at $\mu = 0$

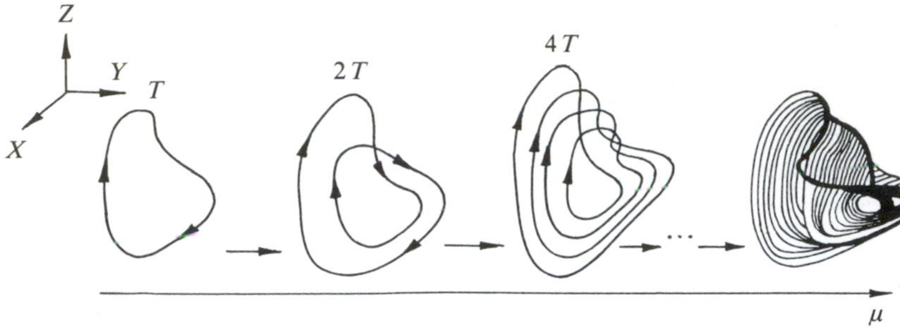

Fig. 3.18. Infinite chain with period doubling leading to a strange attractor (*right-hand side*). The form of the trajectories in phase space is illustrated. The control parameter μ is given on the abscissa. T, 2T and 4T are the corresponding periods of the cycles

3.3.4 Scenarios of the Birth of Strange Attractors

A remarkable achievement in the study of the stochasticity of dissipative systems as pursued in recent years, is the discovery and understanding of the fact that – in spite of the extreme diversity of nonlinear physical systems – there are only a few typical ways in which for small changes of parameters they go over to an irregular behaviour.[2] Surprisingly enough, these transitions are often the same not only for systems of different origin and different complexity, but also for discrete and distributed systems and fields [3.27]. We shall list the most popular scenarios for the onset of chaos in dissipative systems: an infinite chain of period doubling bifurcations [3.28] (Figs. 3.18, 19 and Fig. 6 of Chap. 6 *Nonlinear Physics*); the emergence of stochasticity as stable and unstable periodic motions approach each other to merge and, finally, to vanish (intermittency, see Fig. 3.20) [3.29, 30]; the stochastic destruction of quasi-periodic motions; and the hard onset of stochasticity [3.30, 31].

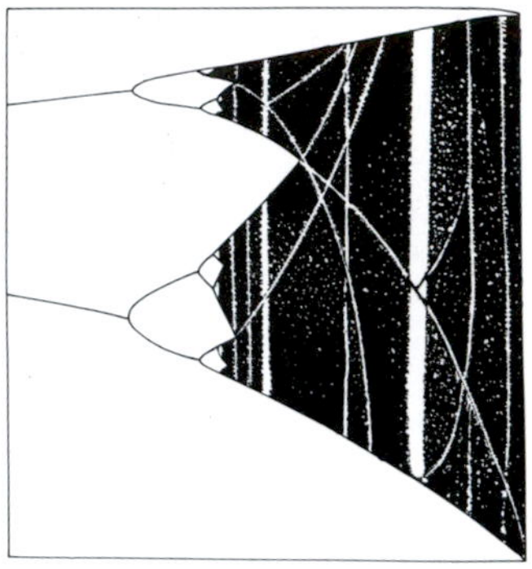

Fig. 3.19. The Feigenbaum doubling tree in the plane "coordinates of cycles versus overcriticality"

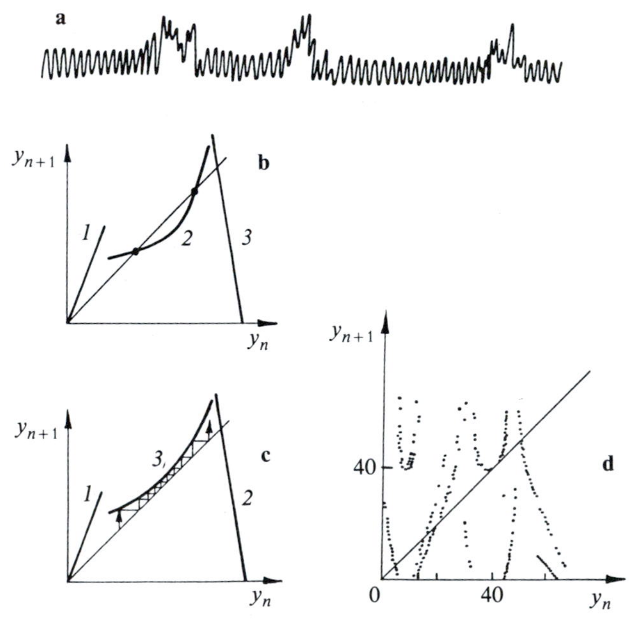

Fig. 3.20a–d. Transition through intermittency:
a) oscillogram showing laminar behaviour with discontinuous turbulent bursts;
b) stable periodic regime (transient chaos);
c) onset of intermittency (stable and unstable cycles merge);
d) intermittency in the Lorenz system

[2] We have already mentioned in Sect. 2.3 that a similar picture holds for periodic self-excited oscillations.

The transition via period doublings is particularly frequent. Its basic features are as follows. Due to the interaction of the oscillations of the basic period the initial limit cycle becomes unstable and in its vicinity a stable cycle with a doubled period is born. There may be an infinite number of such period doubling bifurcations in the bounded space of parameter variations, which, eventually, give rise to a strange attractor in the phase space. For the result of a real experiment see Fig. 6 of Chap. 6 *Nonlinear Physics*.

There may be an infinity of scenarios of bifurcation sequences leading for varying system parameters to the onset of stochastic self-excited oscillations. However, the elementary bifurcations or their sequences containing these scenarios are limited in number. The regimes of soft onset of stochastic self-oscillations are listed in Table 3.1.[3]

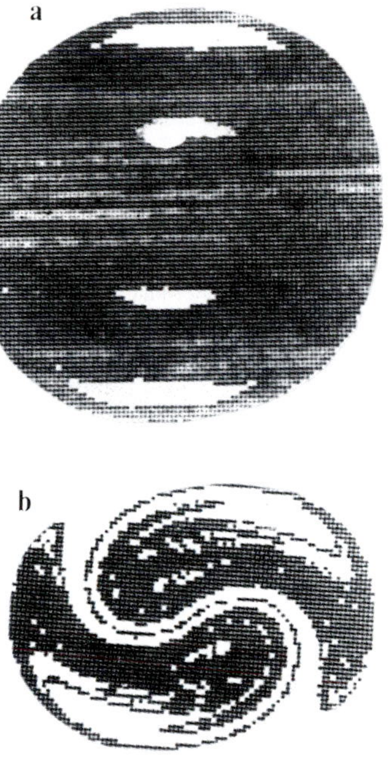

Fig. 3.21a,b. Stochastic manifold of a parametrically excited nonlinear oscillator:
a) without dissipation;
b) strange attractor (with small dissipation)

Characteristic for the hard onset of stochastic oscillations is the transformation of nonattracting homoclinic structures into a strange attractor in the phase space as a result of the loss of stability of simple attractors. For systems close to Hamiltonian ones, a similar situation is observed for increased dissipation, see Fig. 3.21.

In the vicinity of the critical point the paths leading to chaos are for systems of various origins not simply alike, they even satisfy the same universal quantitative relations. Their form depends only on the type of critical behaviour and not on the specific properties of the system (physical, chemical, economical, etc.). We already know some types of critical behaviour – period doubling sequence, intermittency and the transition through quasi-periodicity.

The origin of this universality is related to the closeness of the modelled real systems to the critical point. We shall consider the behaviour of the system having rather similar values of the parameter λ, namely $\lambda_1 = \lambda_{cr} - \varepsilon$ and $\lambda_2 = \lambda_{cr} + \varepsilon$ for $\varepsilon \ll \lambda$. Because of the smooth dependence on this parameter, the respective equations will coincide to an accuracy of $\sim \varepsilon$ on their right-hand sides. For identical initial conditions, the behaviour of the system for $\lambda = \lambda_1$ and $\lambda = \lambda_2$ will deviate only for very large time T, with $T \to \infty$ as $\varepsilon \to 0$. The quantity T is the characteristic time scale at which the difference between the regular dynamics for $\lambda_{cr} - \varepsilon$ and the chaotic dynamics for $\lambda_{cr} + \varepsilon$ will become obvious. This time (at $\varepsilon \to 0$) may arbitrarily exceed all the characteristic times of our dynamic system. Therefore, it is natural to expect that from the point of view of the transition to chaos those details of the behaviour of the particular system which are local in time are rather unimportant. Thus, the transition pattern to chaos must be universal in the vicinity of a critical point. In particular, it follows from this universality that a description of the basic types of critical behaviour may rely on the simplest models demonstrating the basic

[3] A detailed analysis of the emergence of chaos and bifurcation sequences in the vicinity of a critical point, reveals an analogy with the theory of critical phenomena in statistical physics. Within this analogy, phase transitions of the first kind are associated with the hard onset of chaos and phase transitions of the second kind with the soft onset of chaos.

Table 3.1. For different stages of the motion (*first column*) three different scenarios are indicated

Stage	Scenario: Feigenbaum doubling sequence	Intermittency		Ruelle-Takens-Newhouse scenario
Initial dynamics: periodic self-oscillations	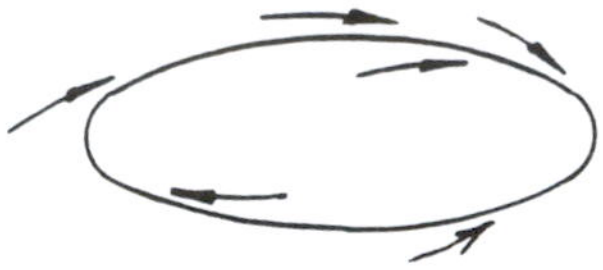			
After the first bifurcation	Period-doubling cycle 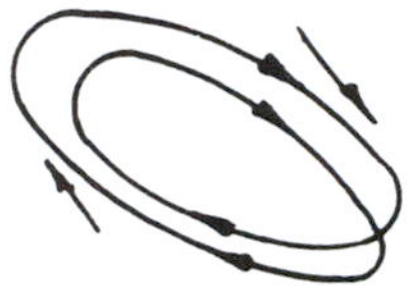	Ergodic torus 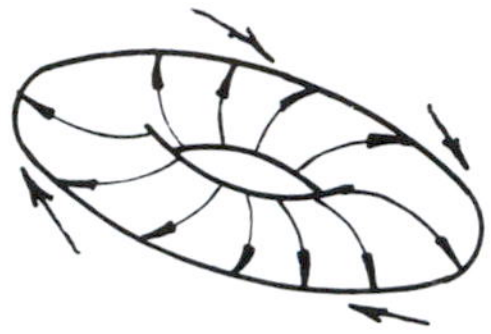		
Precritical behaviour	Traces of two cycles 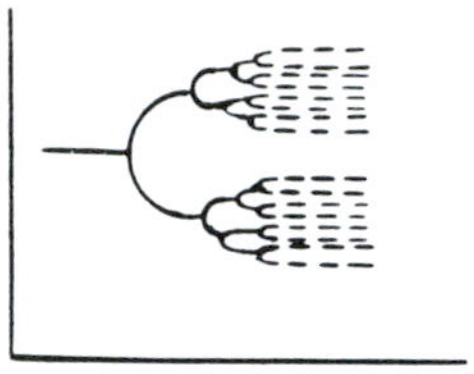 supercriticality →	Synchroni-zation  supercriticality→	Corrugated torus, traces of the secant 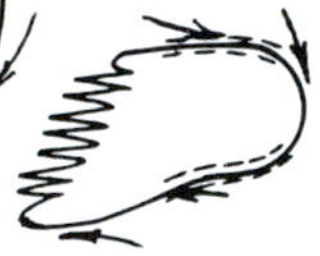	3-D ergodic torus
Past the critical point	Strange attractor 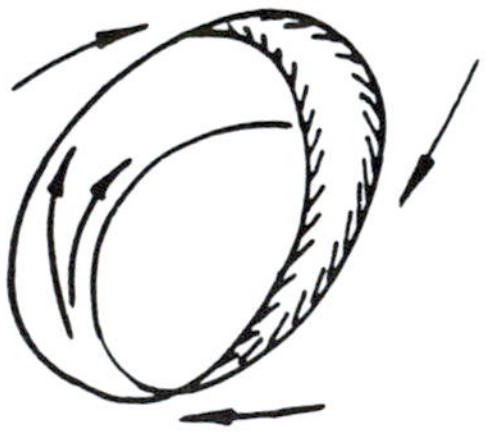	Phase map 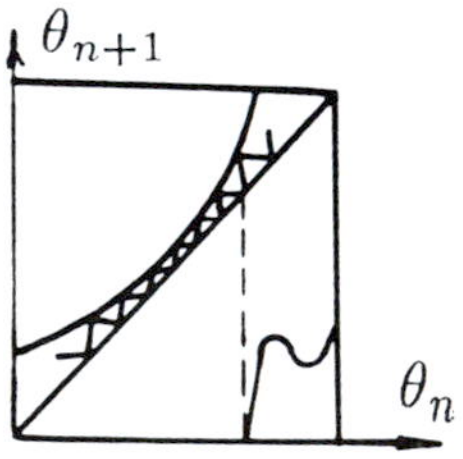		Oscillogram 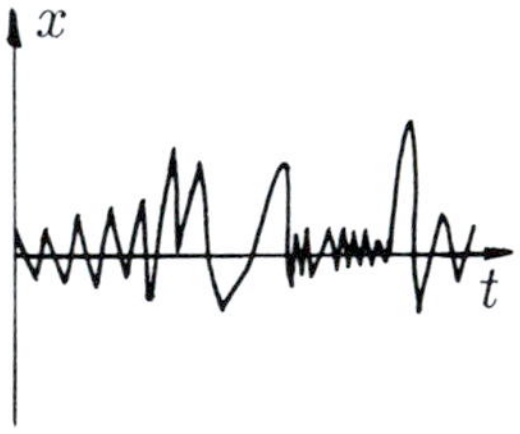
Power spectrum of stochastic self-oscillations	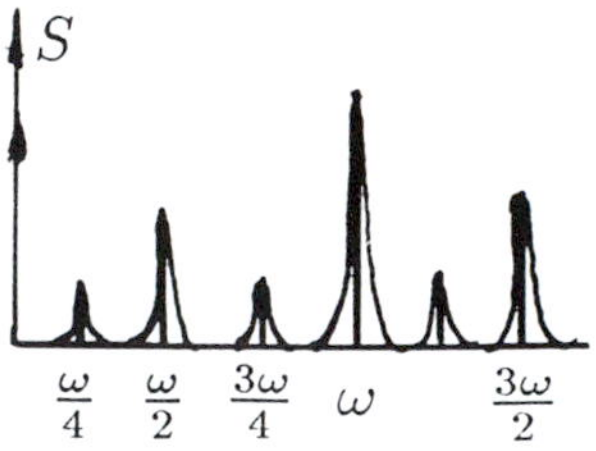	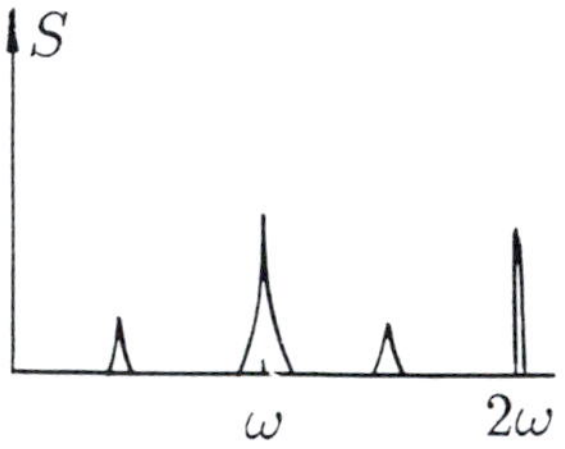		Incommensurable $\omega_1, \omega_2, \omega_3$ 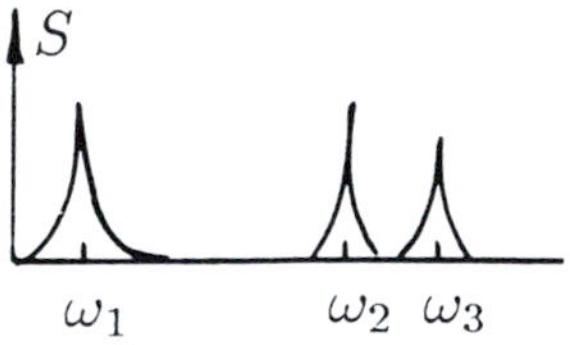

transition paths to chaos. Within these models, it is possible to obtain the qualitative characteristics of the universal behaviour, the similarity constants[4] and critical indices.

3.4 Chaos and Noise

3.4.1 Dimension and Entropy

Stochastic oscillations, like the noise proper, have a continuous Fourier spectrum and a decaying self-correlation function. They can be produced by a dynamic system having a finite number of degrees of freedom; in contrast to random fluctuations or white noise which needs a system in which an infinite number of degrees of freedom can be excited. Evaluating the number of degrees of freedom required for the reproduction of the specific noise under consideration, we can distinguish dynamical chaos from "truly random" motion.

The physical nature of the irregular imbroglio behaviour of a finite-dimensional system is related to the instability of all (or nearly all) individual finite energy motions. Using the phase-space-language, we can explain this as follows: in a bounded region there are unstable trajectories the paths of which diverge due to the instability – and due to the boundedness of the region to which they are confined – and subsequently mix up in a very complicated manner. This complexity of a stochastic set can be described quantitatively.

On a stochastic set we choose an ensemble of sections of trajectories of duration T with the property that they keep the distance ε from each other. Together with their ε-neighbourhood these sections fill the entire set. Therefore, any section of duration T of an arbitrary trajectory on the attractor lies in the ε-neighbourhood of at least one of our sections. We will use $C(T, \varepsilon)$ to denote the number of sections (elements) in the ensemble. The number $C(T, \varepsilon)$ grows with decreasing ε or increasing T. The growth of $C(T, \varepsilon)$ for decreasing ε is quite naturally related to the geometrical complexity of the stochastic set. The growth of $C(T, \varepsilon)$ with increasing T is a consequence of the instability of the trajectories belonging to this set, in particular, to the strange attractor.

To illustrate the above, we shall imagine that we are investigating the attractor with the help of a computer that is able to distinguish the sections of trajectories of duration T only to an accuracy of the order of ε. Our ensemble will then include all possible sections of trajectories of duration T that are stored in the computer memory and $C(T, \varepsilon)$ will denote their number. A decrease in ε will indicate an increase in the sensitivity (accuracy) of our computer and, consequently, the

[4] The states of matter near the critical point are at different temperatures only distinguished by the spatial scale of long-range correlations. Therefore the thermodynamic functions of different states behave in a similar fashion and can be obtained from each other via scaling transformations, a property of critical phenomena named scaling. Regularities of the scaling type are typical for the transition to chaos too [3.32].

increase in $C(T,\varepsilon)$ due to the possible resolution of increasingly finer geometrical structures of the attractor. For a fixed value of ε an increase in T may only for different temporal behaviours of the trajectories of the attractor lead tc a growth of $C(T,\varepsilon)$. This may be related to both the instability of the trajectories and the differences in their behaviour, i.e. to the diversity of the trajectories belonging to the attractor. Instability and diversity of trajectories are described by the topological entropy h_{top} [3.33] that is related to $C(T,\varepsilon)$ via [3.34]

$$h_{\text{top}} = \lim_{\varepsilon\to 0} \overline{\lim_{T\to\infty}} \frac{\ln C(T,\varepsilon)}{T}. \tag{3.7}$$

Another distinguished feature of a stochastic set, besides the diversity of trajectories, is the density or capacity of the trajectories it contains:

$$C = \lim_{T\to\infty} \overline{\lim_{\varepsilon\to 0}} \frac{\ln C(T,\varepsilon)}{-\ln \varepsilon}. \tag{3.8}$$

The quantity C is referred to as its limit capacity or its fractal dimension [3.35]. Let us now introduce the quantity

$$\Theta_{T,\varepsilon} = \frac{\ln C(T,\varepsilon)}{T - \ln \varepsilon} \tag{3.9}$$

known as *Takens'* relation [3.36]. We may use it to give topological entropy and fractal dimension by very much the same expressions, the only difference being the sequence of the limits:

$$C = \lim_{T\to\infty} \overline{\lim_{\varepsilon\to 0}} \Theta_{T,\varepsilon}, \qquad h_{\text{top}} = \lim_{\varepsilon\to 0} \overline{\lim_{T\to\infty}} \Theta_{T,\varepsilon}.$$

We shall now assume that our computer has only two control parameters: accuracy ($\varepsilon \to 0$) and duration ($T \to \infty$). Then, adjusting them in one or the other sequence we shall obtain either dimension or entropy. We can also vary both parameters simultaneously and allow T and $-\ln\varepsilon$ to go to infinity in such a manner that the ratio $T/\ln\varepsilon$ remains constant. In this case, we will obtain new characteristics of the attractor which *Takens* referred to as dynamic invariants [3.36].

In any event, these characteristics of the complex dynamics are related to the instability of individual motions. The instability, in turn, is characterized by the Lyapunov exponents. Therefore it will be natural to find a connection between Lyapunov exponents and dimension. But first let us look a bit closer at the structure of the strange attractor.

A physical theory is not an explanation; it is a system of mathematical propositions whose aim is to represent as simply, as completely, and as exactly as possible a whole group of experimental laws.

Pierre Duhem
"The Aim and Structure of Physical Theory"
translated by P.P. Wiener
(Princeton University, Princeton 1954)
p. ix

3.4.2 The Cantor Structure of a Strange Attractor

The phase spaces of interest may have n dimensions, but to stay with a pictorial example we will consider a three-dimensional phase space. Let us imagine an attractor located in a region bounded by the surface of a two-dimensional torus. We shall consider a beam of trajectories on the way to the attractor (these trajectories describe the transition regimes of motion of the system, that lead to the onset of "stationary" chaos). The trajectories (or, to be more precise, their traces on the intersecting plane) are located in a definite region of the beam's cross-section. Let us follow the changes in magnitude and shape of this region along the beam. We shall take into account that the element of space is stretching along one direction (transverse) in the neighbourhood of the saddle trajectory and contracting into the other. Since the system is dissipative, the contraction will be more pronounced than the stretching and the volumes will become smaller. The directions must change along the trajectories, otherwise they would go to infinity. The resulting beam cross-section will be smaller in area and will acquire a flattened and at the same time bent shape. The cross-section of the beam will, eventually, be divided into a system of bands embedded into each other, see Fig. 3.22. As time progresses, the number of these bands will rapidly grow (along the beam of trajectories) while their widths will become narrower. The attractor emerging in the limit $t \to \infty$ contains an infinite number of layers that do not touch one another – surfaces on which the saddle trajectories are located (we have already established, see Fig. 3.3, that these trajectories have the attracting directions "outwards" the attractor). With their ends these layers are interconnected in a complicated manner; nearly each trajectory belonging to the attractor walks over all the layers and, in a sufficiently large time, it will approach any point of the attractor (ergodicity). Overall volume of layers and total area of their cross-sections are zero. Such sets are in one direction Cantor sets. The Cantor properties of the structure are the most characteristic features of the attractor. These findings hold also in the more general case of an n-dimensional ($n > 3$) phase space.

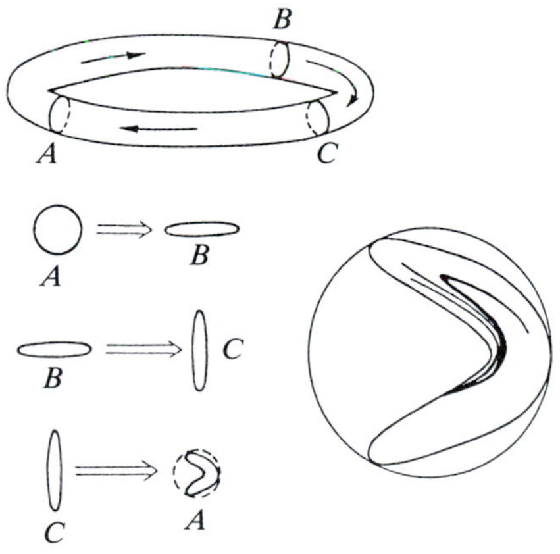

Fig. 3.22. Cantor structure of strange attractor. The cross-section A of the flow of the trajectory transforms into cross-section B which goes over into C which in turn becomes the new A, etc. thus giving rise to the "cantorian" cross-section of the attractor

The volume of the strange attractor in the initial phase space is always zero. But it will be nonzero in a phase space with lower dimension. The latter is determined as follows. We shall divide the entire n-dimensional space into small cubes having length ε and volume ε^n. Let $N(\varepsilon)$ be a minimal number of cubes that cover the attractor completely. We shall define the dimension D of the attractor as the limit

$$D = \lim_{\varepsilon \to 0} \frac{\ln N(\varepsilon)}{\ln(1/\varepsilon)}. \tag{3.10}$$

Not just in this context is the notion of the dimension crucial: Mathematical solitons as analytical solutions of (partial differential) equations were first discovered in $1 + 1 = x+t$ dimensions and readily extended to $2+1$ dimensions, but $3+1$ dimensions and higher look like hard rock.

The dimension of the set defined in such a natural fashion was found [3.35] to coincide with the limit capacity or fractal dimension specified by (3.8). The existence of the limit (3.10) indicates the finiteness of the attractor volume in the D-dimensional space: for small ε we have $N(\varepsilon) \approx V/\varepsilon^D$ (where V is a constant); whence it follows that $N(\varepsilon)$ can be considered as the number of D-dimensional cubes covering the volume V in the D-dimensional space. The dimension determined according to (3.10) can apparently not exceed the overall dimension n of the phase space, but it can be smaller than n and, unlike the conventional dimension, it may be fractional: that is the case for Cantor sets.[5]

We will draw your attention to the following important fact. For the motion established on the attractor, the energy dissipation is on the average compensated for by the energy supplied from the source of the system's nonequilibrium. Consequently, if we follow the time evolution of the "volume" element belonging to the attractor (in a certain phase space its dimension will be determined by the attractor dimension), then this volume will be conserved on the average, – its contraction in some directions will be compensated for by a stretching into others caused by the divergence of close trajectories. We can employ this property to estimate the attractor's dimensions by means of a different method.

3.4.3 Dimension and Lyapunov Exponent

The average characteristics of attractors with ergodic motion can be found by analysing the motion of a single unstable trajectory belonging to the attractor. An individual trajectory reproduces all the properties of the attractor, provided we follow it for an infinitely long time.

Let $\boldsymbol{X} = \boldsymbol{X}_0(t)$ be the equation of such a trajectory, one of the solutions of the basic nonlinear equations

$$\dot{\boldsymbol{X}} = \boldsymbol{F}(\boldsymbol{X}, \mu). \tag{3.11}$$

We shall consider the deformation of the "spherical" volume element in its motion along this trajectory. The deformation determined by (3.11) is linearized over the difference $\boldsymbol{\xi} = \boldsymbol{X}(t) - \boldsymbol{X}_0(t)$, i.e. over the deviation of the trajectories next to the given one. Written in components, these equations have the form

$$\dot{\xi}_i = A_{ik}(t)\xi_k, \qquad A_{ik}(t) = \left.\frac{\partial F_i}{\partial X_k}\right|_{\boldsymbol{X}=\boldsymbol{X}_0(t)}. \tag{3.12}$$

[5] The n-dimensional cubes covering the set may be "nearly empty", therefore D may be smaller than n. For conventional sets definition (3.10) gives obvious results. Thus we obtain for a set of isolated points $N(\varepsilon) = N$ and $D = 0$; for the section l of a line, $N(\varepsilon) = l/\varepsilon$ and $D = 1$; for the area S of a two-dimensional surface, $N(\varepsilon) = S/\varepsilon^2$, $D = 2$, etc.

In its motion along the trajectory the volume element contracts in some directions and stretches into others and the sphere transforms into an ellipsoid. During the motion the directions of the semi-axes of the ellipsoid change as well as their lengths. We shall express the lengths through $l_j(t)$ where j labels the direction. The limits

$$\lambda_j = \lim_{t\to\infty} \frac{1}{t} \frac{\ln\, l_j(t)}{\ln\, l(0)} , \tag{3.13}$$

(where $l(0)$ is the radius of the original sphere at the time $t = 0$) are referred to as the Lyapunov characteristic exponents. The quantities determined in this manner are real numbers whose value equals the dimension of the (n-dimensional) space. The one out of these numbers that corresponds to the direction along the trajectory itself, equals zero.[6]

The rate at which the largest eigenvalue grows is called the Lyapunov exponent of the trajectory. The growth of these eigenvalues is crucial for judging the long-term behavior of the dynamical system. In particular, we find instabilities when some of the eigenvalues grow exponentially with time.

M.C. Gutzwiller
"Chaos in Classical and Quantum Mechanics"
(Springer, New York, Berlin, Heidelberg 1990) p. 17

The sum of the Lyapunov exponents determines the variation (average along the trajectory) of the elementary volume in phase space. The local change of the volume at each point along the trajectory is specified by the divergence $\mathrm{div}\dot{\boldsymbol{X}} = \mathrm{div}\dot{\boldsymbol{\xi}} = A_{ii}(t)$. We can show that the average value of the divergence along the trajectory is [3.37]

$$\lim_{t\to\infty} \frac{1}{t} \int_0^t \mathrm{div}\dot{\boldsymbol{\xi}} dt = \sum_{j=1}^{n} \lambda_j . \tag{3.14}$$

For a dissipative system this sum is always negative, i.e. arbitrary volumes contract in n-dimensional phase space. We will arrange the Lyapunov exponents in the order

$$\lambda_1 \geq \lambda_2 \geq \ldots \geq \lambda_k \geq 0 \geq \lambda_{k+1} \geq \ldots \geq \lambda_n$$

and take into account the stable directions to the extent needed to compensate for the stretching caused by the contraction. If the resulting sum $\sum_{j=1}^{m} \lambda_j$ were exactly equal to zero, then the integral value m would be the attractor dimension determined via the Lyapunov exponents. However, usually the sum of the integral number of exponents is nonzero and we must take into account a fraction of the next $m+1$ "contracting" exponent to meet the requirement of the "conservation of the phase volume on the attractor". Thus, the attractor dimension D_λ will be somewhere between m and $m+1$, where m is the number of exponents in the given sequence the sum of which is still positive, but upon adding λ_{m+1} it becomes negative.[7] The fractional portion

[6] The solution of (3.12) (with initial conditions for $t = 0$) describes, actually, the neighbouring trajectory only as long as the distances $l_j(t)$ are small. However, (3.13) remains valid for arbitrarily large times. In this equation there appears the relative length changes for large t; within the linear approximation, this coincides with the result of successive relative variations in time intervals during which the equations may be linearized.

[7] Inclusion of the Lyapunov exponent equal to zero, introduces the contribution +1, corresponding to the dimension along the trajectory itself, to the dimension D_λ. That explains why the dimensions of the stochastic manifold on the Poincaré plane and in the initial phase space differ by unity.

$d < 1$ of the dimension $D_\lambda = m + d$ can be found from the equation

$$\sum_{j=1}^{m} \lambda_j + d\lambda_{m+1} = 0 \tag{3.15}$$

(*F. Ledrappier* [3.38]). Since only the least stable directions are taken into account in the calculation of d (the negative exponents of λ_i having the highest absolute values at the end of the sequence corresponding to fast motions are omitted) the estimate of the dimension obtained through D_λ is, generally speaking, an estimate for the upper limit $D_\lambda \gtrsim C$ [3.39].

If the attractor dimension is $D_\lambda \gtrsim 2$, then the phase trajectories forming the attractor are located in a thin layer near some surface. This quasi-two-dimensionality of the attractor allows for an approximate description of its motion using a one-dimensional Poincaré mapping, which connects the coordinates of preceding and following intersections (by a section plane, see Fig. 3.7) of the trajectory belonging to the attractor. This family of attractors with $|D - 2| \ll 1$ includes, in particular, the Lorenz attractor and many others.

It now appears quite natural that a system of arbitrary dimension should give rise to such transitions to stochasticity like those observed in one-dimensional maps like period doubling, intermittency, etc. – provided the dimension of its stochastic set is in the vicinity of the point where chaos is born only a little higher than 2 (the "soft" onset of the attractor). Thus, we have shown that one-dimensional maps provide indeed a good model for studying attractors with dimensions close to two.

The characteristics of the dimension of stochastic motion are extremely important from different viewpoints. On the one hand, they allow us to increase our knowledge of randomness; on the other hand, they are quite handy in purely applied problems connected with signal processing (their coding, identification, and so on). Indeed, within a traditional analysis of a random signal (spectral, correlation) we can hardly say anything about its source. In particular, we don't know whether it is an ordinary noise signal (that cannot be reproduced by an algorithm) or whether it is produced by a deterministic system, even if it is a very complicated one. This problem can be solved by determining the signal dimension (see below). The finite dimension D_λ, or C indicates that the signal can, in principle, be reproduced by means of a dynamic system of an order not higher than $2D_\lambda + 1$. Thus, the value of the dimension D_λ gives an estimate for the number of the degrees of freedom in the system (medium) that take part in the formation of the stochastic signal of interest. As the dimension of the time series (tending to infinity) increases, the chaotic signal approaches more and more a truly random signal. From this point of view, the source that is traditionally taken to be random, can be considered as the motion of a dynamic system on a strange attractor of infinite dimension. The dimensions for different types of dynamic motion are listed in Table 3.2.

The main goal of physics is to describe a maximum of phenomena with a minimum of variables.

CERN Courier

Table 3.2. From the *left* to the *right* the respective dimensions D, entropies H, signals $U(t)$ and types of attractors are shown

Dimension	Entropy	Signal $U(t)$	Type of attractor	
$D = 0$	$H = 0$	$u(t) = \text{const}$ 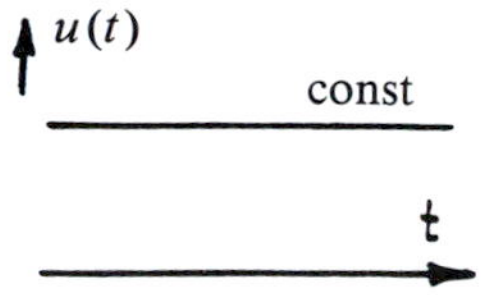	Stable equilibrium	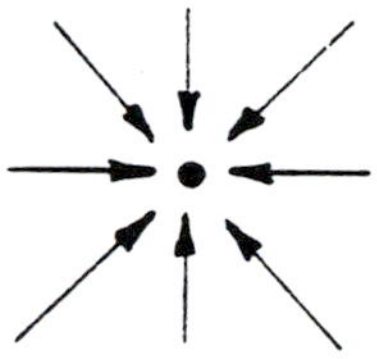
$D = 1$	$H = 0$	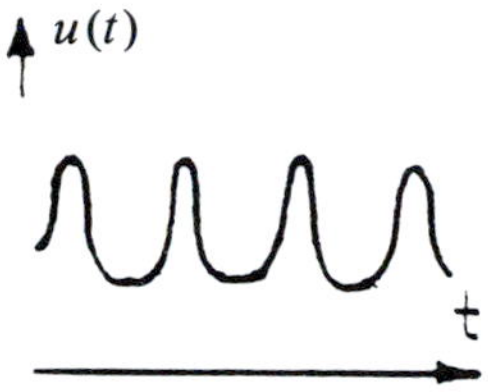	Limit cycle periodic self-excited oscillations	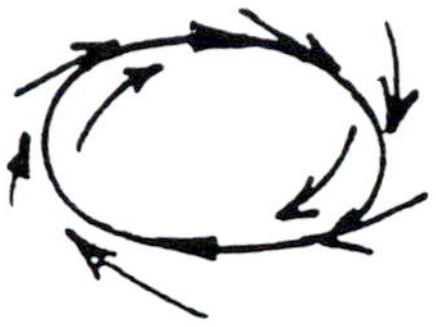
$D = 2$	$H = 0$	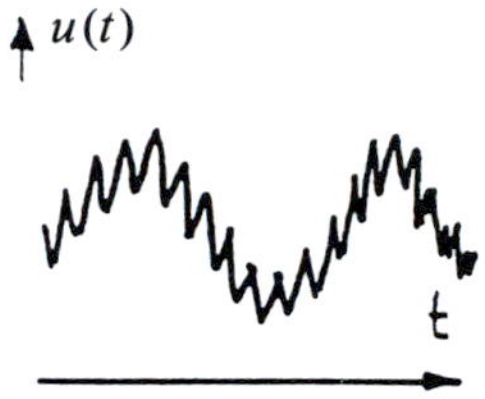	Open winding motion on a 2-d torus. multi-periodic self-excited oscillations	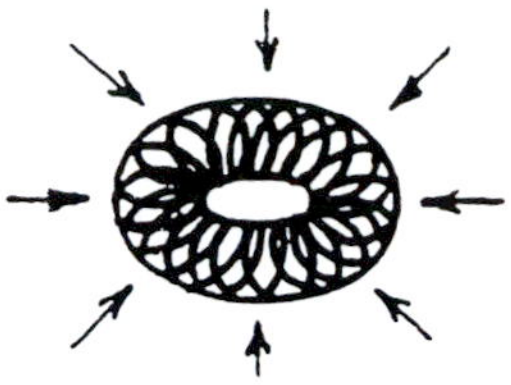
$D = 2.06$	$H > 0$	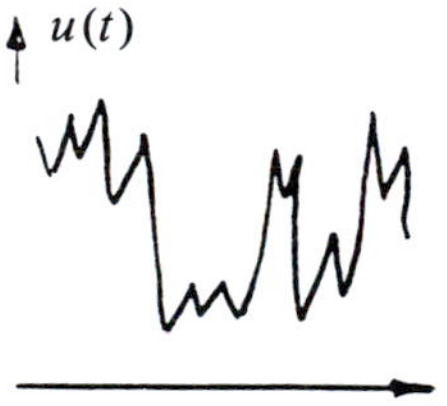	Strange attractor (Lorenz attractor) Stochastic self-excited oscillations	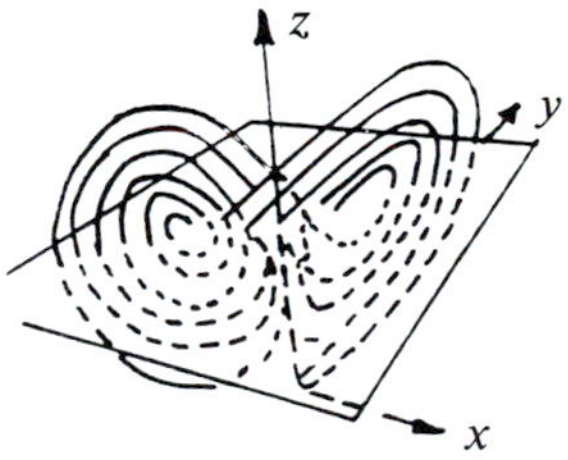
$D \gg 1$ $D \to n$ $n \to \infty$	$H > 0$		Multi-dimensional attractor	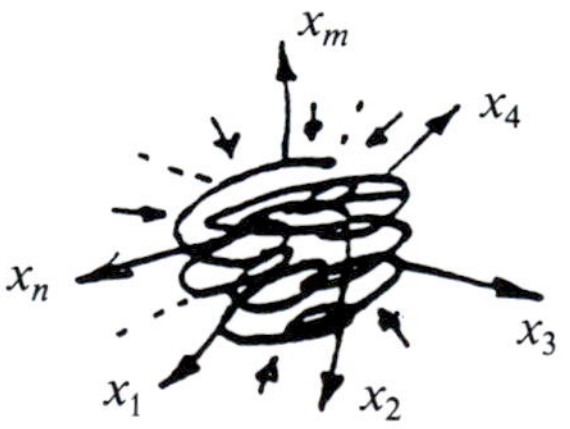

3.4.4 Deterministically Generated and Random Signals

The idea of processing random signals to reproduce the properties of the sources generating the received signals was proposed relatively recently by *Takens* [3.35]. He reasoned: if the signal is generated by a finite-dimensional dynamic system then it will be possible first to reproduce the corresponding limit set (a strange attractor, in particular) in a certain effective space and then to determine on this set the characteristics of motion, such as entropy and dimension. An essential contribution to the development of these ideas and their practical implementation was made in 1983 by *Grassberger* and *Procaccia* [3.40]. They proposed not to reconstruct the limit sets in phase space, but, instead, to treat a specific time series (of sufficiently long duration) of the physical quantity studied.

We usually have a single observable, for example, one component of the velocity field that is measured as a function of time at one point of a hydrodynamic flow. Since the dimension of the effective phase space, into which the stochastic set corresponding to this time series is embedded, has the dimension n, we need to know n independent functions of time $u_k(t)$. Takens proposed to obtain them as follows: the observable is considered at discrete instants of time $t, t+\tau, \dots, t+(m-1)\tau$; the respective values of $u_k(t)$ yield the coordinate of the point at the moment t in the m-dimensional space. As t varies, we shall obtain in this space a trajectory reproducing a certain set. The dimension C_m (or D_λ^m) of this set, its entropy, etc. can be calculated from (3.8) or (3.15). The next step is the analysis of the m-dependence of C_m. Apparently, for small m the dimension C_m will increase with m. In the case of a noise signal, the growth will proceed without saturation. But if the signal is produced by a dynamic system, then the growth will cease at some $m = M$, see Fig. 3.23. The value C_M is the dimension of the reproduced limiting set, in particular, of the strange attractor.

It should be noted that it is not convenient to calculate the dimension of the attractor from (3.8) or (3.10) since it is not clear, for instance, up to which limit ε should be decreased. Recent investigations [3.40–43] have shown that a more effective method is to use the correlation integral. It is approximately given by[8]

$$C^m(\varepsilon) = \left\{ \frac{1}{N} \sum_i^N \left[\frac{1}{N} \sum_j^N \Theta(\|\boldsymbol{u}_i - \boldsymbol{u}_j\| - \varepsilon) \right] \right\}$$
$$\equiv \frac{1}{N} \sum_i^N \frac{N_i(\varepsilon)}{N} = \frac{N_\varepsilon}{N} . \qquad (3.16)$$

Here $\boldsymbol{u}_j = (u_j, u_{j+1}, \dots, u_{j+m-1})$ is a point on the trajectory in m-dimensional space, N the total number of points of the observable within the considered time interval, Θ the Heaviside function, and $\|\dots\|$ denotes the distance between a pair of neighbouring points.

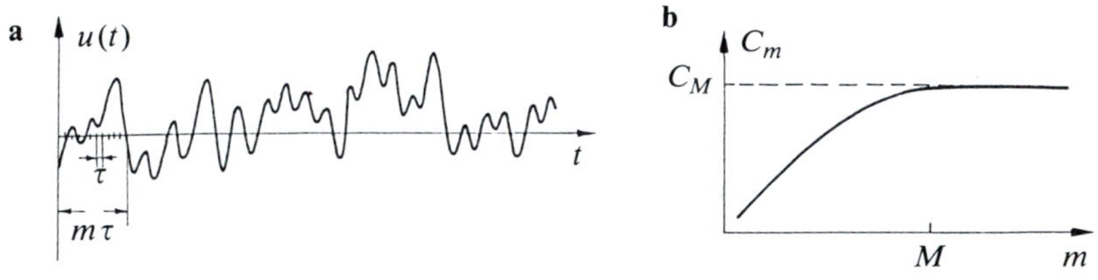

Fig. 3.23a,b. From the experimental time signal the dimension of the corresponding stochastic set is extracted: *a*) discretization of the time series; *b*) saturation of the dimension C_m of the set in the constructed (C_m, m) phase space

[8] A rigorous determination of the correlation integral implies the transition to the limit $N \to \infty$.

Thus, the correlation integral is the average number of pairs of points in the time series, with the distance between them in the space u_j being smaller than ε. Grassberger ascertained that for small ε, the correlation integral depends on ε like

$$C^m(\varepsilon) = \varepsilon^\nu \cdot \exp[-Km\tau] \;, \tag{3.17}$$

where K is the Kolmogorov entropy. To this precision we can easily establish the equality between the fractal dimension C introduced earlier and the correlation dimension $\nu = C_m$:

$$C = \lim_{\varepsilon\to 0} \frac{\ln N(\varepsilon)}{\ln(1/\varepsilon)} = \lim_{\varepsilon\to 0}\left(\nu + \frac{\ln N - Km\tau}{\ln(1/\varepsilon)}\right) = \nu = C_m \;. \tag{3.18}$$

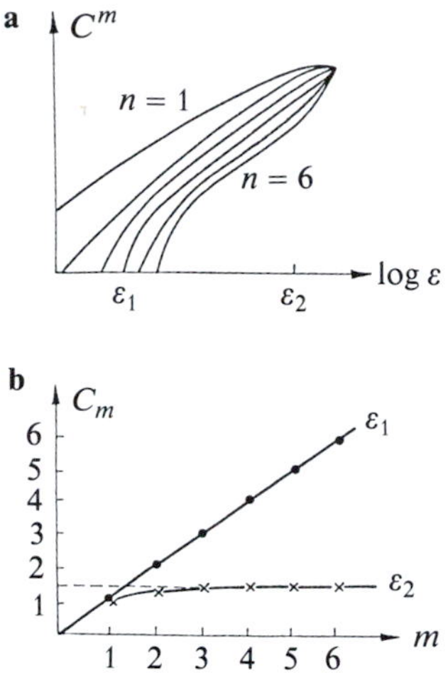

Fig. 3.24a,b. Correlation integral and dimension:
a) The correlation integral C^m is seen to have different slopes for different intervals of ε.
b) The correlation dimension C_m for a Hénon map with noise

In the treatment of experimental data, the dimension is usually calculated approximately, directly by the slope of the plot of $\ln C^m$ vs. $\ln\varepsilon$. These quantities behave differently for signals of different origins, which allows to analyse signals of unknown origin, and in some instances, to separate the dynamical component (having a finite dimension) of the signal from the noise.

In an attempt to determine the slope of the plot $\ln C^m(\varepsilon) = f(\ln\varepsilon)$ we encounter the problem that the plot has different slopes for different ε-intervals, see Fig. 3.24a. Which of the slopes should be considered as the "true" dimension? Before attempting to answer that question, let us first ponder about the possible causes of the bends in our plot. We can say right away that they may be accounted for by either purely technical peculiarities in the treatment of the time series (its insufficient duration, for instance) or by principal features of the dynamic system that has produced this time series.

If, for example, the attractor is inhomogeneous in the effective phase space (i.e. the image point is more frequently encountered in some portions of it than in others), then the value ν will be different for different ε.[9] However, for sufficiently large N the density of the points on the attractor will properly represent its structure. Thus, it is the attractor's "inhomogeneity" that leads to the bends in the plot, which can in principle be disposed of. We should emphasize that the required duration of the time series is related to the measured dimension. The higher the dimension, the greater a duration of the time series is needed to provide a sufficient filling of the attractor by the individual points. An empirical estimate [3.43] yields: $\ln N \sim \nu\ln(\varepsilon_{\max,\min})$, where $(\varepsilon_{\max}, \varepsilon_{\min})$ denotes the interval ε in which $C^m(\varepsilon) \sim \varepsilon^\nu$, see (3.17).

The strong bends in the correlation integral are, most often, due to the "structured" nature of the observable, to the fact that the signal contains components possessing different dimensions. The components may originate from different systems, including noise generators (e.g. a communication channel with noise). In the simplest case, the structured signal has the form $u(t) = u_0(t) + \delta_1 u_1(t) + \ldots + \delta_k u_k(t)$, where

[9] It may even happen that a short time series simply does not allow for the time to manifest some individual details of the attractor.

$u_i(t)$ is generated by the dynamic system whose attractor dimension equals ν_i with $\delta_1 > \delta_2 > \ldots > \delta_k$ and $\nu_0 < \nu_1 < \ldots < \nu_k$. Then, the components with increasingly lower amplitudes δ_i will become more apparent in the correlation integral, as ε decreases successively. Consequently, the plot will give rise to the intervals $(\varepsilon_i', \varepsilon_i'')$ with the tangent of the angle of incidence being $\sim \nu_i$. It seems to be natural to refer to the number of bends $(K-1)$ for such a signal as to its degree of "structuredness". It should be emphasized that an increase in the duration of the time series does not remove the bends, but makes them more pronounced.

Since any observable corresponding to a real process is associated with noise, a dynamic system reproducing exactly this observable must be infinite-dimensional. If the signal-to-noise ratio is not too small, then the correlation integral for such a realistic signal will inevitably have a bend separating the scale of the dynamical component from the small scales of ε where the dimension is not determined because of the presence of noise. Thus, when analysing the local slopes of the plot $\ln C^m(\varepsilon)$ vs. $\ln \varepsilon$, we now will be able to distinguish a chaotic signal of dynamic origin from additional white noise. An illustrative example is shown in Fig. 3.24b, where the correlation integral and the local slopes are given at different ε for the Hénon mapping with noise.[10]

[10] The nice book by *Parker and Chua* [3.44] may be helpful for the reader if he wants to get more acquainted with the techniques of obtaining the dimensions of complex signals.

4. Structures

One-dimensional structures like solitons and saw-toothed shock waves are extremely interesting and their study is very instructive. But when we come closer to reality and investigate two- or three-dimensional systems, a new, wonderful world comes to light. It is almost as fascinating as Nature itself.

4.1 Order and Disorder – Examples

Our surprise with regard to stochastic motions arising in simple nonlinear systems is very much the same as that experienced in the twenties and thirties when the "spontaneous" emergence of periodic motions was observed in dissipative systems (like a nonlinear amplifier coupled with an LC-circuit). Even today, having acquired much more experience, we find the onset of regular oscillations out of initial disorder (or fluctuations) in the absence of an organizing force not at all trivial. Indeed, even if the evolving instability is intrinsically resonant it is not strongly periodic disturbances that arise but an entire spectrum of such disturbances. What are the mechanisms for the selection of a specific, regular perturbation out of arbitrary disturbances? What peculiarities must be inherent to such an "anti-entropic" system which transforms the energy of an external aperiodic and, in general, irregular source into periodic motion?

Essentially the same problems as in the classical theory of nonlinear oscillations are posed nowadays in connection with quite peculiar new phenomena. An example is the emergence of spatial order out of initial disorder and the spontaneous formation of complicated spatial structures in one-dimensional nonequilibrium media. A keen interest in such phenomena arose in the fifties and sixties and was associated with the problems of chemical kinetics and biology. In particular, waves in a cardiac muscle (1954) [4.1] and in a model of morphogenesis (1952) [4.2] as well as periodic oscillations in an autocatalytic chemical reaction [4.3, 4] were described qualitatively. Almost in the same period, a theory of regular spatial structures in certain hydrodynamic flows (like the Bénard cells in thermoconvection, see, e.g. [4.5] and the Taylor vortices between rotating cylinders [4.6]) was constructed. The onset of complicated ordered structures in nonlinear media or in spatial ensembles of different origins were soon found to be described by similar models and solutions, see [4.7–9]. This enabled the researchers (not

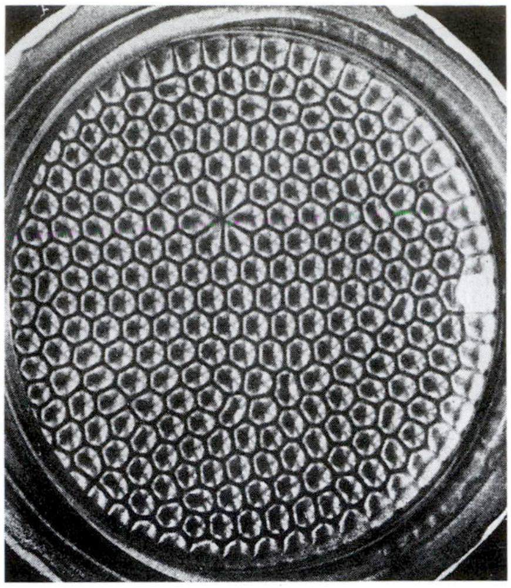

Fig. 4.1. Hexagonal prismatic cells (Bénard cells) arising in a fluid layer heated from below

for the first time in nonlinear physics and in the theory of oscillations and waves) to apply the experience and knowledge gained in the study of a particular problem (like the investigation of the propagation of a flame) to the analysis of another one (like the evolution of populations in ecological problems or the propagation of a disturbance in cardiac tissue). This situation facilitated the evolution of new ideas and concepts (dissipative structures, auto-waves, reverberators, and so on), and notions leading to the construction of basic universal models describing formation and existence of structures. A new branch of nonlinear sciences referred to as nonequilibrium thermodynamics [4.7], synergetics [4.8] or as the theory of self-organization [4.10, 11] has actually appeared.

The formation of the structure of hexagonal prismatic cells (Bénard cells, see Fig. 4.1) in a fluid layer heated from below is a traditional physical example of self-organization. The necessary conditions for the emergence of such a structure are the nonequilibrium and the dissipation of the underlying nonlinear system. A developing convective instability leads to an increase of the velocity and temperature field perturbations in a certain spatial interval, then, due to the effect of scale competition (which manifests itself only in the presence of dissipation), a lattice with a specific scale is seen to survive. We will show below that the hexagons are formed as a result of the phase synchronization of lattices having different spatial orientations, see Fig. 4.2.[1] Experiments show that neither the lattice scale nor the cell structure depend on the conditions in the periphery of the layer, provided its horizontal size is great enough.

Synergetics deals with the profound analogies between the self-organized behavior of seemingly quite different systems in physics, chemistry, biology, sociology and other fields.

H. Haken
"Synergetics"
(Springer, Berlin, Heidelberg, New York 1978)

Self-organization has been observed in biological systems in relatively simple situations, for instance, in a homogeneous ensemble of amoeba-like cells [4.12]. These cells secrete a special hormone.

[1] Such a synchronization is possible in fluids where viscosity (surface tension or other diffusion constants) are temperature dependent, see the text below.

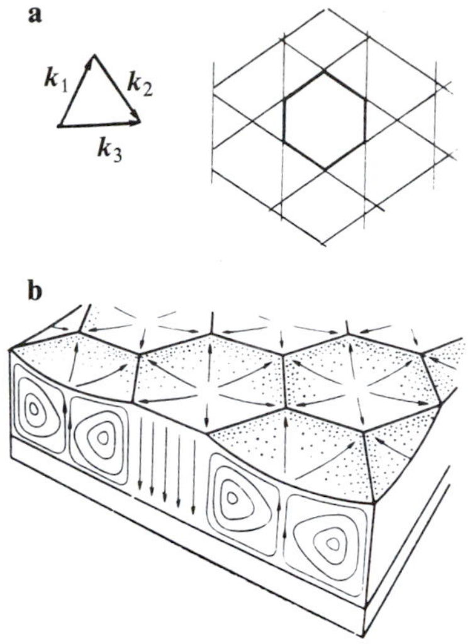

Fig. 4.2a,b. The hexagonal prismatic cells that arise in the superposition of three spatial lattices with the same periods:
a) tilted against each other at an angle of 60°;
b) current lines in a liquid layer being in the regime of Bénard convection

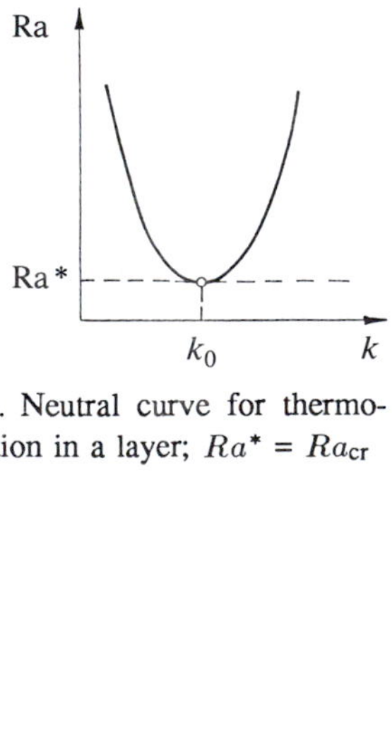

Fig. 4.3. Neutral curve for thermoconvection in a layer; $Ra^* = Ra_{cr}$

If there is enough food then the other cells do not respond to this "message" and all cells live independently. Under more specific conditions, the following occurs. One of the cells speeds up the secretion of the hormone and synchronizes the hormone secretion of its nearest neighbours, which, in turn, synchronize their neighbours, and so on. Having been excited by the hormone, the cell moves towards the exciter. Thus, two counter-propagating motions – the diverging waves of the stimulator, or synchronization, and the motion of the cells towards each other – occur, which lead to the formation of a new organism.

Now we shall come back to convection and consider in more detail the mechanism of the formation of the Bénard cells, whose shape and scale are within finite limits independent of initial and, which is of particular importance, boundary conditions. We have already said, that in this case there are two decisive effects – mutual synchronization and mode competition – that were first investigated in detail by A.A. Andronov in the thirties. We shall take the simple example of cellular convection in silicon oil (see Fig. 4.2) for which the temperature dependence of viscosity, $\nu(T)$, is significant. Convective motions with a characteristic scale (Fig. 4.3) arise in the oil layer with small excess over the instability threshold. The growth of perturbations with the wave vector $\boldsymbol{k}_{01}$ gives rise to the simplest spatial structure in the form of convective rolls. However, for temperature-dependent viscosity,[2] this structure is unstable with regard to the excitation of modes having a different vector orientation $\boldsymbol{k}$ like rolls located perpendicular to the initial ones (their coexistence results in the formation of rectangular structures).[3] The dependence $\nu(T)$ can usually be considered to be quadratic, then a resonant coupling arises between the three modes having equal scales, $\boldsymbol{k}_{01} \pm \boldsymbol{k}_{02} = \pm \boldsymbol{k}_{03}$, see Fig. 4.2a. The superposition of these modes having equal amplitudes and phases synchronized in space,

$$v_z(x, y) \sim \cos\left[\frac{k_0}{2}x\right] \cos\left[\frac{k_0 x + \sqrt{3}k_0 y}{4}\right] \cos\left[\frac{k_0 x - \sqrt{3}k_0 y}{4}\right]$$

(with the vertical component v_z of the fluid velocity) corresponds to the nontrivial structures in the form of hexagonal Bénard cells – the fluid rises in the centre of the cell and drops down near its boundaries (or vice versa, when $\partial\nu/\partial T > 0$). The orientation of the cells in space is arbitrary and depends on the initial conditions. The competition of the modes of different scales ensures the stability of the structure with regard to the onset of other structures. As the excitation amplitude grows, the elementary cells of the lattice (like solitons) gain significant independence, and the spatio-temporal description is then more natural than the "mode" description. A little later we will use the spatio-temporal description to analyse an ensemble of "self-structures".

---

[2] The temperature dependences of surface tension and of other dissipative parameters lead to similar effects.

[3] Note that rectangular structures are formed only for a strong temperature dependence of the viscosity.

We consider another example to demonstrate the effect of mode synchronization in the onset of ordered structures in a nonequilibrium medium. It relates to the emergence of dissipative solitons, in particular, in a long LC-transmission line with tunnel diodes, see Fig. 2.29. In this case the one-dimensional waves are described by

$$\partial_t u + v_0 \partial_x u + \beta \partial_{(2n+1)x} u = \alpha u^2 - \nu_1 u + \nu_2 \partial_{xx} u \ . \tag{4.1}$$

Here β describes dispersion, $\nu_{1,2}$ stand for low- and high-frequency dissipation, respectively, and α is the active nonlinearity. For strong dispersion, the evolution of the perturbations in such a medium can be described by a few modes having, for example, the frequencies ω or 2ω. The equations for their amplitudes and phases will be analogous to (2.5), to an accuracy of the terms responsible for linear dissipation:

$$\begin{aligned} &\dot{A}_1 = A_1 A_2 \cos\Phi - \delta_1 A_1 \ ; \quad \dot{A}_2 = 0.5 A_1^2 \cos\Phi - \delta_2 A_2 \ , \\ &\dot{\Phi} = -\left(2A_2 + \frac{A_1^2}{2A_2}\right)\sin\Phi \quad \text{with} \quad \Phi = \varphi_2 - 2\varphi_1 \ . \end{aligned} \tag{4.2}$$

The only significant differences are the equal signs on the right-hand sides of the equations for A_1 and A_2. The related physics is that the harmonics are damped or built-up simultaneously, i.e. the waves exchange energy with the nonequilibrium medium, rather than with one another. When exceeding the instability threshold at a favourable phase difference ($\Phi = 0, \pi$) the amplitudes of the harmonics tend within this model in a finite time (or within a finite distance) to infinity: $A_{1,2} \sim 1/(t^0 - t)$ with $t^0 = 1/A_{1,2}(0)$. In other words, an explosive instability occurs. It is essential that the explosive instability is accompanied by a fast phase synchronization of the interacting waves [4.13]. This synchronization leads to the onset of strongly nonlinear waves, in particular, to dissipative solitons,

$$u(x,t) = (3\nu_1/\alpha)\cosh^{-2}\left[\sqrt{\nu_1/2\nu_2}(x - v_0 t)\right]$$

which were also observed experimentally, see Fig. 2.29b. This occurs when a large number of harmonics interact in a dispersionless medium ($\beta = 0$).[4] Unlike the "conservative" solitons referred to in Chap. 2, dissipative solitons propagate only with the speed of linear perturbations v_0. The phase portrait (4.1) for $\beta = 0$ for the stationary waves $u = u(x - v_0 t)$ coincides with Fig. 2.29 b.[5] The mode synchronization in non-one-dimensional nonequilibrium media results in much more complicated wave structures. A flowing liquid film is an example of such a medium. We can write an approximate equation for the deviation u of the film surface from the unperturbed level [4.14]:

It will turn out that equations governing self-organization are intrinsically nonlinear. From those equations we shall find in the following that, modes may either compete, so that only one survives, or coexist by stabilizing each other.

H. Haken
"Synergetics"
(Springer, Berlin, Heidelberg, New York 1978) p.14

---

[4] In this case the number of interacting modes is limited by high-frequency damping.

[5] The fact that the phase portrait of the self-oscillating system coincides for stationary waves with the phase portrait of a conservative oscillator appears at first sight to be paradox. The explanation is that all perturbations move in such a medium with the speed v_0.

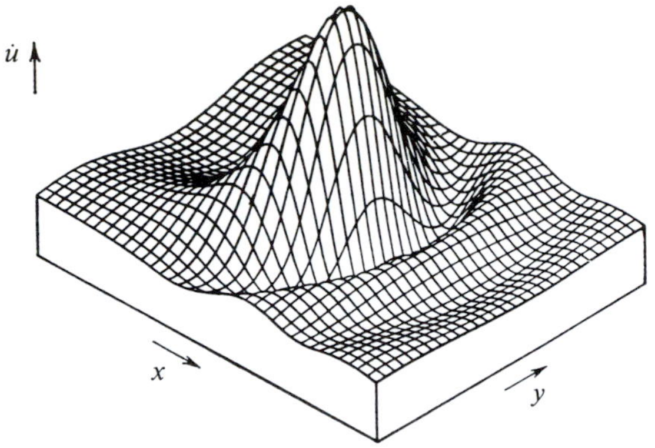

Fig. 4.4. Horseshoe-shaped soliton obtained in solving (4.3)

$$\partial_t u + 4u\partial_x u + \partial_{xx} u + \Delta_\perp^2 u - \gamma \partial_{yy} u = 0 \tag{4.3}$$

(here $\gamma > 0$, and $\Delta_\perp^2 = \partial_{xx} + \partial_{yy}$). The numerical solution to this equation is presented in Fig. 4.4 [4.15] for $u(x \to \pm\infty) = 0$; this is a localized structure in the form of a horse-shoe with an oscillating forefront and a monotonically decaying rear – a two-dimensional analogue of a stationary shock wave. This solution gives a good description foı waves observed experimentally on a film flowing down a surface, see Fig. 4.5.

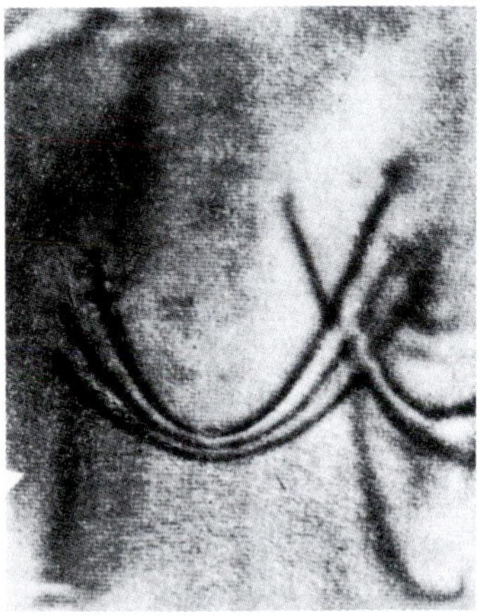
Fig. 4.5. Waves flowing downwards on a film

Examples show that self-organization is the result of developing spatially inhomogeneous instabilities with their subsequent stabilization due to the balance between internal dissipative consumption and the energy supply (mass, etc.) from the source of nonequilibrium. Very much like the onset of self-excited oscillations in a generator! But in an LC-generator with discrete parameters "self-organization" is realized only in time, while it occurs in this particular case both in time and space. Therefore, it is not surprising that the examples of self-oscillations in continuous systems (the generation of pulses in a laser [4.16] being the most popular among them) are simultaneously examples of self-organization phenomena. Bearing in mind our desire to unify the problems under consideration and taking into account that self-organization phenomena include both the formation of static structures and propagating fronts (moving structures), that is "nonoscillatory" processes, we employ from now on the term "self-organization" rather than "self-oscillations".

Since self-organization occurs in space and time it is extremely rich in form. There are dissipative structures, including localized ones, self-structures, solitary fronts (combustion waves [4.17], population waves), pulses (in nerve fibers [4.18]), guiding centres and reverberators (in cardiac tissue [4.19]), depression waves (in the brain [4.20]), and others.

The formation of structures against a background of random initial noise is, at first sight, a phenomenon opposed to the birth of chaos in a deterministic system. However, both chaos and order are manifestations of nonlinearity. But chaos is observed in systems with unstable behaviour of most individual motions while order is inherent to systems with stable behaviour. As we have shown in the previous chapter, the transition from stable to unstable motion may occur even at very small variations of the parameters of the dynamic system, in the transition through the critical point. Note that the established chaos bears the imprints of the vanished order, even rather far away from the transition boundary. Examples of "stochastized" ordered structures are the coherent structures in turbulent flows [4.21, 22], the "chemical" turbulence in the form of an ensemble of concentric or spiral waves (see Plates 36 and 37 of Chap. 16 *Nonlinear Physics*) [4.23, 24], the Langmuir wave turbulence in the form of a soliton "gas" [4.25], etc. To view structures or order as the antipode of chaos would be too strong. It appears much more natural to attempt to describe different levels of order or different degrees of chaos in nonequilibrium media. These problems are discussed in here and in the chapter to follow.

If we create a universe, let it not be abstract or vague but rather let it concretely represent recognizable things. Let us construct a two-dimensional universe out of an infinitely large number of identical but distinctly recognizable components. It could be a a universe of stones, stars, plants, animals or people.

M.C. Escher
in "The World of M.C. Escher"
J.L. Locher (Ed.)
(Abrams, New York 1971) p. 40

4.2 Attractors and Spatial Patterns

4.2.1 Examples of Equations

In spite of the enormous diversity of stable spatial structures – crystalline and quasi-crystalline patterns, lattices with various defects, fractal dendrites, particle-like formations, etc. – the unity of the dynamical mechanisms of their birth and evolution in nonequilibrium media of different origins, as discovered relatively recently, allows us to formulate some basic equations of the nonlinear theory of such structures.

For not too large supercriticalities there are many instances in which the motion of the medium (field) can be assumed to be given along one of the spatial coordinates. For the description of the field dynamics with regard to the two other spatial variables, we can derive the so-called amplitude equations. It is natural to classify these equations into two groups: gradient equations having only static solutions as $t \to \infty$ and oscillatory equations that allow for periodic, quasi-periodic or chaotic limiting solutions when $t \to \infty$.

The simplest and in this sense most canonical gradient equations are: the nonlinear diffusion equation describing the dynamics of a real field

$$\partial_t u = \mu u - u^3 + \Delta u \ , \tag{4.4}$$

a similar equation for a complex field

$$\partial_t u = \mu u - |u|^2 u + \Delta u \ , \tag{4.5}$$

Swift-Hohenberg's model [4.26] taking into account diffusion dispersion

$$\partial_t u = \left[\mu - \left(q^2 + \nabla^2\right)^2\right] u - u^3, \tag{4.6}$$

and *Haken's* equation [4.27]

$$\partial_t u = \left[\mu - \left(q^2 + \nabla^2\right)^2\right] u + \beta u^2 - u^3, \tag{4.7}$$

in which the additional quadratic nonlinearity, responsible, in particular, for the formation of polyhedral lattices in two-dimensional media, is taken into account phenomenologically. Equations (4.4–7) supplemented by boundary conditions, in particular by periodic ones like

$$u(x, y, t) = u(x + L, y, t) = u(x, y + L, t) \tag{4.8}$$

or by demanding that the solutions vanish at the boundary

$$u(x, y)\big|_\Gamma = \nabla u(x, y)\big|_\Gamma = 0 \ , \tag{4.9}$$

can (for a complex field) be written in the gradient form

$$\partial_t u = -\frac{\delta F}{\delta u} \quad \text{or} \quad \partial_t u = -\frac{\delta F}{\delta u^*} \ . \tag{4.10}$$

Here $\delta F/\delta u$ is the variational derivative of the functional $F\{u, u^*\}$, which represents the free energy. For example, for the system (4.6), (4.9), F has the form

$$F = \int \left(-\frac{\mu}{2}u^2 + \frac{1}{4}u^4 + \frac{1}{2}\left[\left(q^2 + \nabla^2\right) u\right]^2\right) dx\, dy \ . \tag{4.11}$$

Anyone who plunges into infinity, ... must divide his universe into distances of a given length, into compartments recurring in an endless series. Each time he passes a borderline between one compartment and the next, his clock ticks.

M.C. Escher
ibid. p. 40

The zero variation of F,

$$\delta F \Big|_{u=u^0(x,y)} \delta u = 0 \ ,$$

corresponds to the equilibrium state (4.10), while the sign of the derivative dF/dt determines the stability of the equilibrium. According to (4.10, 11),

$$\frac{dF}{dt} = -\int |\partial_t u|^2 dx\, dy \lesssim 0 \ , \tag{4.12}$$

dF/dt is negative for any nonstationary solution, therefore only static states are stable in gradient systems. The spatial patterns corresponding to these states may be diverse and, in the general case, form a continuous set to which a continuum of trivial attractors – equilibrium states – corresponds in the phase space.

As the supercriticality increases, the temporal behaviour of the spatial structures in nonequilibrium media becomes significantly more complicated and they can no longer be described by gradient models. We shall give examples of such "basic" equations, that describe not only static, but also pulsating spatial patterns as $t \to \infty$. Some of them are the *generalized Ginzburg-Landau equation* (GGLE) [4.28–30]

$$\partial_t u = \mu u - (1 + i\beta)|u|^2 u + \kappa(1 - ic)\Delta u \ , \tag{4.13}$$

the parametric analogue of which is [4.31]

$$\partial_t u_1 = -i\mu u_2^* - (1+i\beta)|u_1|^2 u_1 + (1-ic)\Delta u_1 \ ,$$
$$\partial_t u_2 = -i\mu u_1^* - (1+i\beta)|u_2|^2 u_2 + (1-ic)\Delta u_2 \ , \tag{4.14}$$

the nonautonomous *sine-Gordon equation* [4.32]

$$\partial_{tt} u - \Delta u + \sin u = F(x,y)\sin\omega t - \nu\partial_t u \ , \tag{4.15}$$

and some other models that are not so frequent in nonlinear physics.

As compared to gradient systems, the space of solutions of systems (4.13–15) (with the boundary conditions (4.8) or (4.9)) is in general more sophisticated. It can already contain both stable periodic and quasi-periodic solutions, corresponding to attractors in the form of ergodic winding on a two- or three-dimensional torus, as well as the solutions that are stochastic in time (and space, for example, when $L \to \infty$). We will see that the strange attractors may correspond to purely temporal chaos in an ensemble containing a small number of regular spatial structures. In this case the spatial pattern is relatively simple and is repeated in different time intervals (this being the only manifestation of chaotic dynamics). In this case, however, a radically different situation is possible – the complicated spatial patterns corresponding to the solution on a strange attractor do not repeat themselves in time and, under definite conditions (if the medium can be assumed to be unbounded), can be considered to be disordered in time and space. This is exactly what we refer to as spatio-temporal chaos, see the chapter on "turbulence", as well as [4.31, 32].

Softness triumphs over hardness, feebleness over strength. What is more malleable is always superior over that which is immoveable. This is the principle of controlling things by going along with them, of mastery through adaptation.

Lao Tzu

4.2.2 Multistability. Defects

The simplest spatial patterns emerging in homogeneous dissipative media near the threshold of instability are regular lattices either in the form of a sequence of parallel rolls or of square or hexagonal cells. A minimal number of modes (eigenfunctions) of the linear problem usually correspond to such collective excitations of the medium. They are one-dimensional standing waves for rolls, the superposition of mutually orthogonal standing waves of equal wavelengths for squares, and for hexahedrons they are three phase synchronized standing waves with an angle of 60° with respect to one another. It is rather unproblematic to describe these structures, but pondering about the ways in which they arise rather difficult questions come to mind. For example: How is a particular lattice chosen from random initial conditions? What explains the resulting orientation of the lattice? Why is it that in some cases identical lattices with different orientations coexist and in other situations they don't? We come across many questions of this kind.

As an example for discussing these problems let us use model (4.7). It describes, in particular, Bénard-Marangoni convection in a plane layer of fluid heated from below [4.33]. We will now assume periodic boundary conditions. In the system of interest the limiting regimes are always static. This is implied by the fact that for (4.7, 8) or (4.9) there exists a Lyapunov (free energy) functional in the form

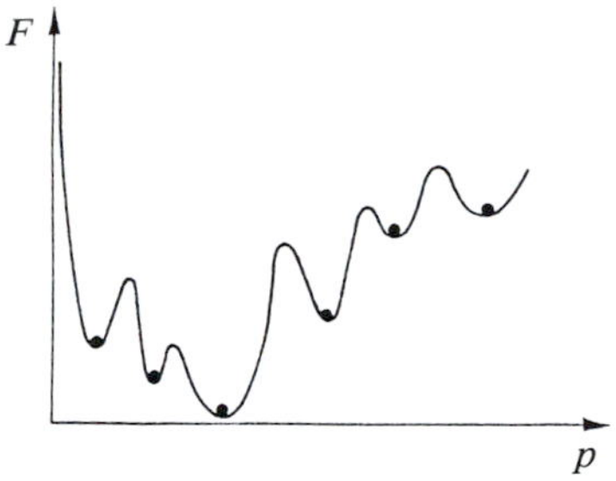

Fig. 4.6. Dependence of the potential function F on the coordinate p in configuration space

$$F(t) = \int \left(-\frac{\mu}{2}u^2 - \frac{\beta}{3}u^3 + \frac{1}{4}u^4 + \frac{1}{2}\left[\left(q^2 + \nabla^2\right)u\right]^2 \right) dx\, dy . \tag{4.16}$$

The analysis of this functional shows that for not too small μ it has a large number of local minima corresponding to different spatial patterns, see Fig. 4.6. The absolute minimum corresponds to a regular lattice of rolls (if $\beta < \beta^*$) or of hexahedrons (if $\beta > \beta^*$). The mentioned local minima give rise to lattices having different defects, see Figs. 4.7, 8. The phenomenon of multistability is understood to be typical for the considered class of nonequilibrium media. The initial conditions determine the specific pattern to emerge for $t \to \infty$. The transition from one stable pattern to another is governed by finite fluctuations.

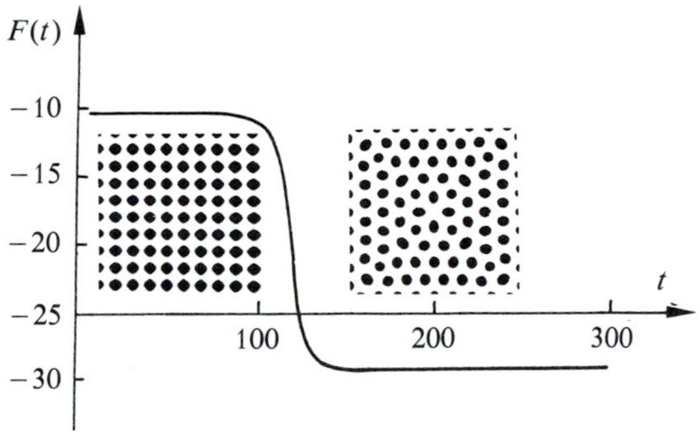

Fig. 4.7. Transition from a metastable lattice to a stable one (F(t) is the free energy functional)

One of the principal mechanisms of the onset of lattices is the competition of structures discussed above. In the simplest case of regular structures, ordinary mode competition describes the interaction of the given spatial distributions in time. The competition of the spatially homogeneous modes appears to be the simplest tool for the explanation of the onset of a stable lattice in the form of n parallel rolls emerging from a given initial lattice containing, say, $(n+1)$ rolls. However, such a spatially homogeneous mode suppression has been observed neither in the numerical experiments with system (4.7, 8), nor in the physical systems with weakly supercritical convection [4.5]. Actually, the most energetically profitable process in the interaction of similar modes is the gradual restructuring of the lattice period in the largest portion of space. It is accompanied by a local symmetry breaking implying the emergence of defects; after their dissapearance a stable periodic lattice emerges, see Figs. 4.9, 10. We will be able to construct a theory for such nonstationary defects, if we assume that the amplitudes of the interacting modes are not only functions of time, but also of the spatial coordinates. With such a description we can, in particular, explain the effect of changing spatial symmetries due to the birth and disappearance of defects, as well as the coexistence of different spatial patterns [4.34]. The boundary between different stable lattices may be viewed as an extended defect with the respective separatrix in the space of stationary states, see also [4.35].

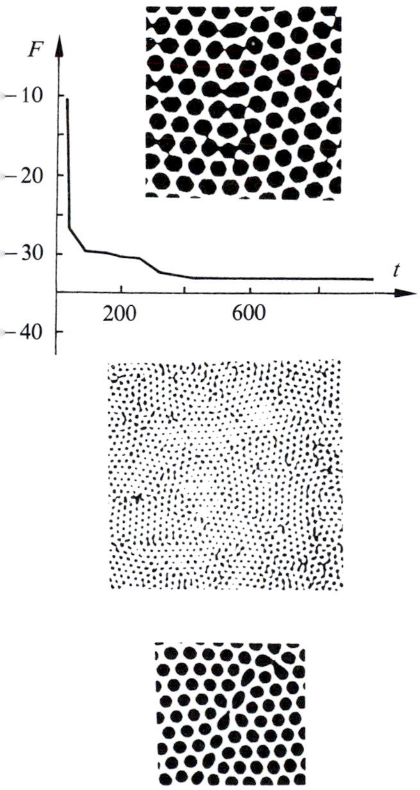

Fig. 4.8. Emergence of a stable lattice with defects

Fig. 4.9

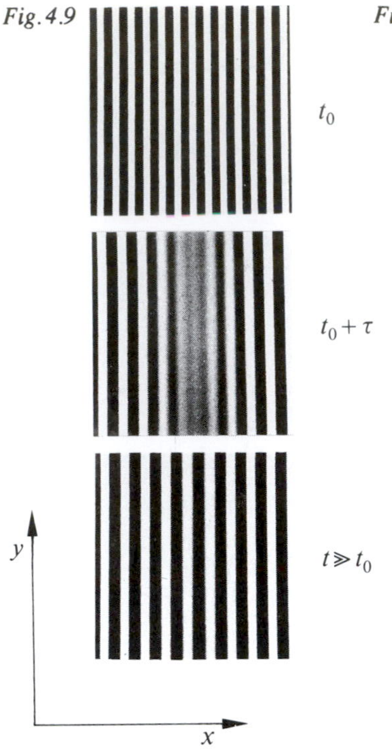

Fig. 4.10

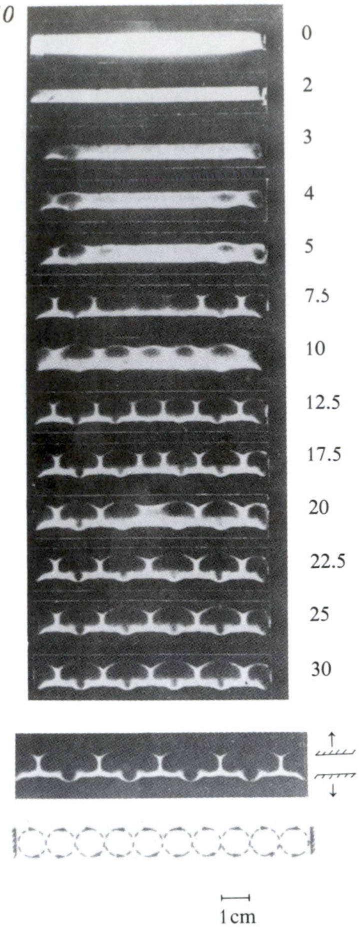

Fig. 4.9. Spatio-temporal competition of convection rolls for different times t (computer experiment)

Fig. 4.10. Experimentally observed rearrangement of a lattice with N = 6 rolls into a lattice with N = 5 rolls. On the *right-hand side* the time t is given in minutes; ↑ means "top" and ↓ "bottom"; *at the very bottom* of the figure the positions of the side walls are indicated

Apparently, the paths followed in the formation of stable regular structures vary and depend on initial conditions. Figure 4.11 presents the results of a numerical experiment, which demonstrate different types of pattern formation for a hexagonal lattice. All these routes correspond to the transition of the system to the same state, having the lowest free energy ($F_{\min} = -38$, for the parameters given in the figure), but the state emerges in different ways, see Fig. 4.11. In some cases, for example, through the emergence of a honey-comb lattice from a standing wave or from a (nearly) point perturbation (Fig. 4.11) the lattice is formed due to the successive development of instabilities, first, of the primary one and then of modulation instability [4.36].

Changing or violating the symmetry of two-dimensional lattices implies the formation of defects; there are two groups: global ones (leading to the changes in the field pattern throughout the entire space) and point (or local) violations corresponding to local symmetry violation (along one or two coordinates, respectively) of initially homogeneous lattices.

Defects of the first group are caused by interactions like collective excitations, having incommensurate scales. It can be described in terms of nonlinear mode dynamics, see e.g. [4.37, 38] and Fig. 4.12. Defects of the second group are due to topological peculiarities of a two-

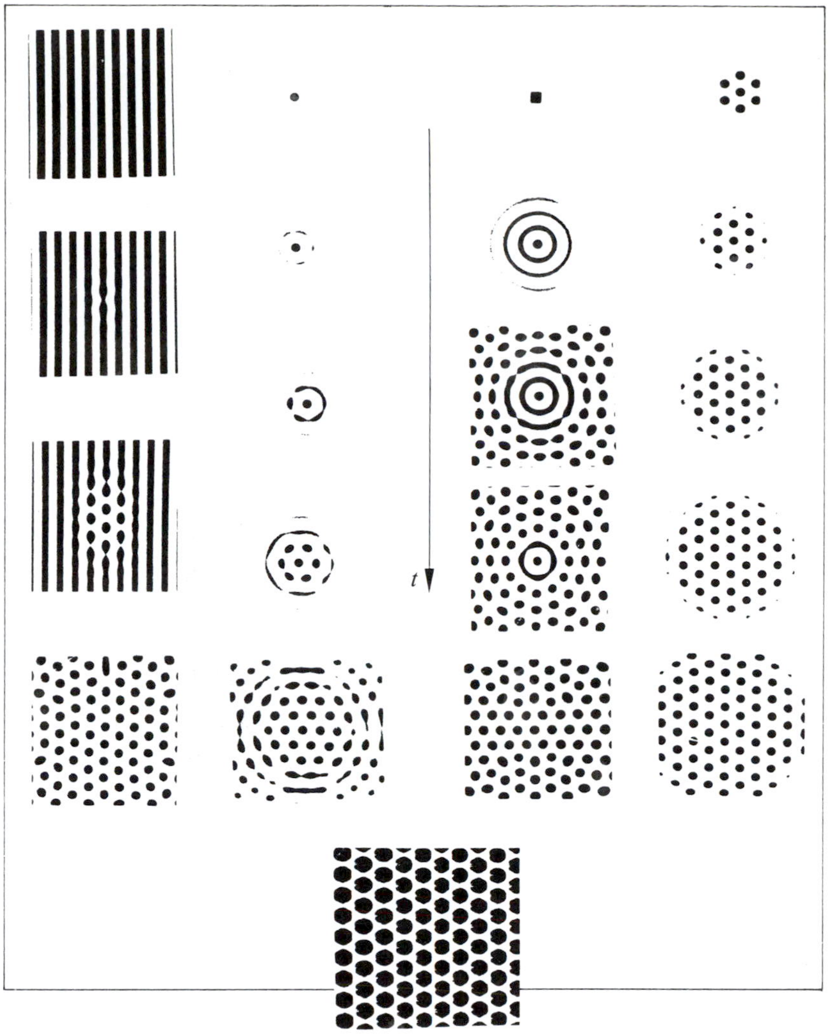

Fig. 4.11. Different paths for the birth of hexagonal lattices ($\beta = 1.5$; $q = 1$)

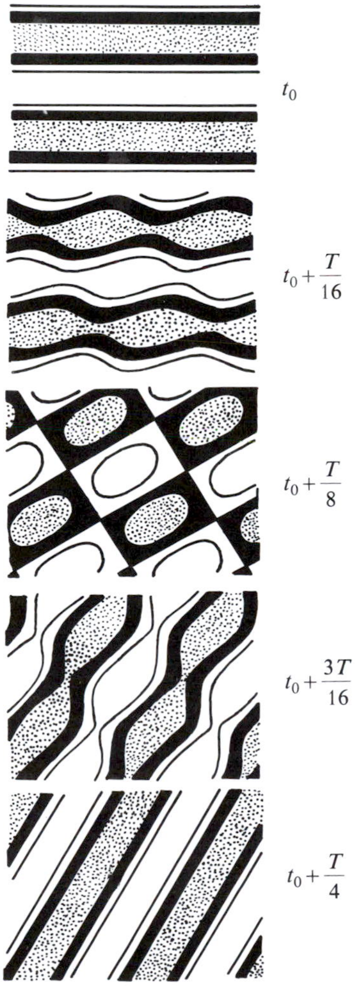

Fig. 4.12. Global transformations of a lattice as the time t increases (T is the time period)

dimensional field and are created and destroyed as the equilibrium states (or their continua) appear and vanish on the phase plane as a result of local bifurcations.

The static, in particular, the topological features of the defects emerging against the background of regular lattices in nonequilibrium media (for example, in different liquid flows, see Fig. 4.13) do not differ from the corresponding characteristics of defects, for instance, in the theory of crystals, where they have been investigated in ample detail [4.39]. The peculiarity of nonequilibrium systems is that the topological field defects move, arise and vanish continuously, i.e. for fields in nonequilibrium media the nonlinear dynamics of defects is of primary interest. In particular, turbulence can be interpreted as the chaotic dynamics of defects, provided the supercriticalities are not too large, see also Chap. 5.

The dynamics of defects has been investigated most thoroughly for potential systems. The motion of defects in such systems is always completed by the generation of one of the static states, to which the local minimum of the functional F corresponds (the defects either

a

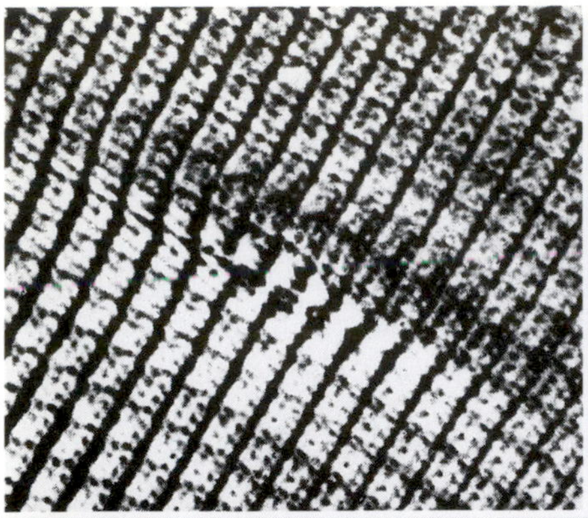

b

Fig. 4.13a–e. Defects in real hydrodynamical flows:

a) dislocation of capillary waves or Faraday ripples;

b) dislocation of modulation waves on a background of Faraday ripples;

c) dislocated pairs (defects) in electrohydrodynamic convection (in planar aligned nematic liquid crystals [courtesy of I. Rehberg]);

d) temporal dynamics of convective defects (in units of τ);

c

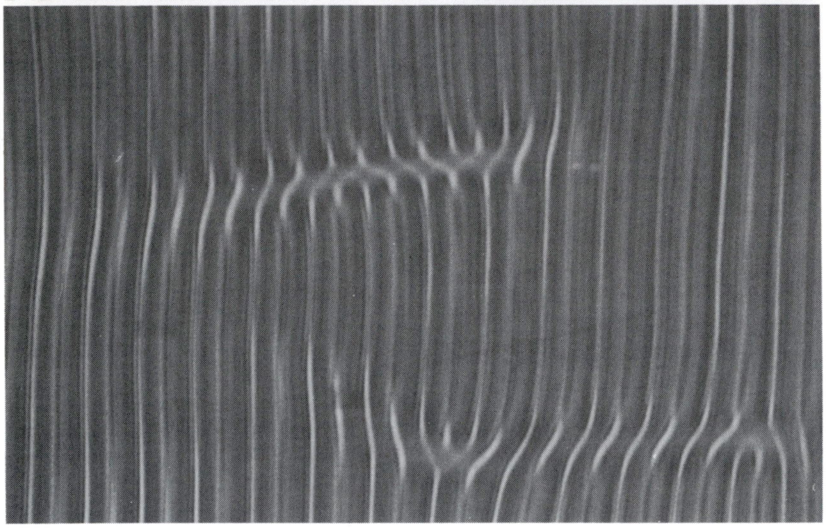

d

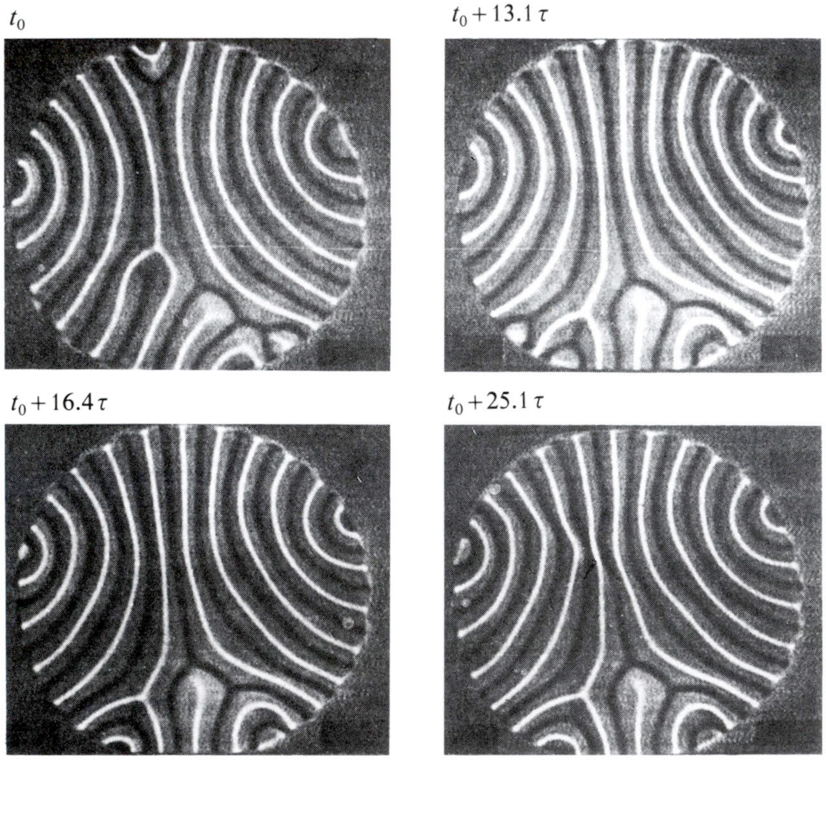

e) increasing number of defects observed in liquid crystals as supercriticality increases [courtesy of I. Rehberg]

e

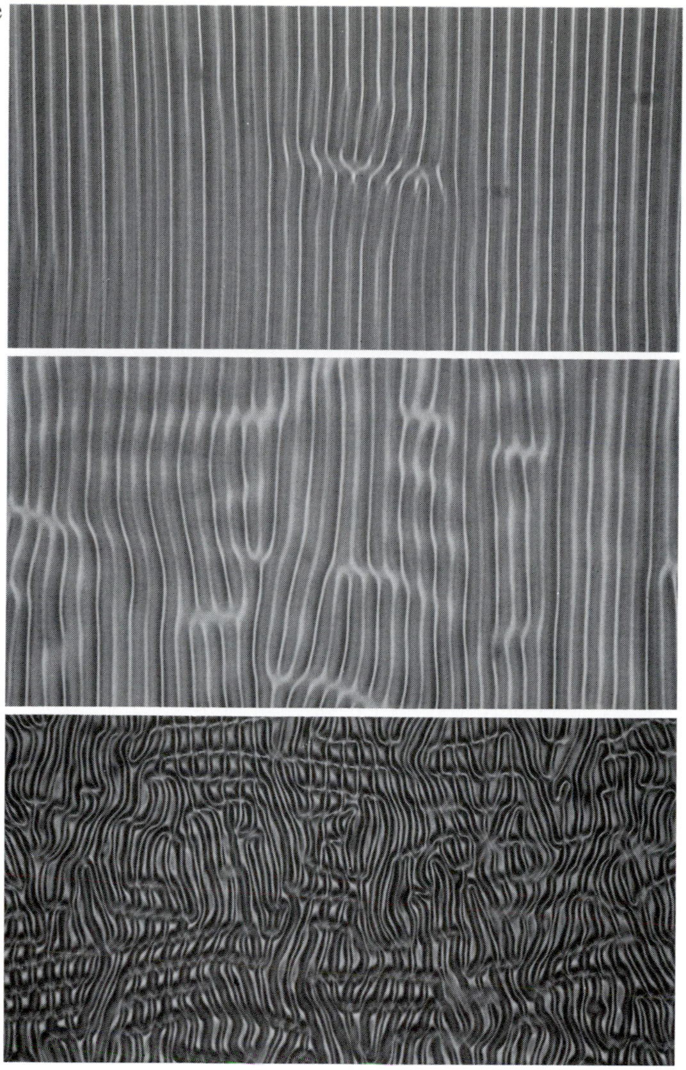

come to a halt or vanish). In such a case the change in the free energy F can be understood as the work done by some force, which, by analogy with the force acting on the dislocation in a crystal under the action of pressure, is referred to as the Peach-Köhler's force [4.33]. When the roll pattern with the wave vector $2(n+1)\pi/Lx)$ moves by dy and is replaced by rolls with the wave vector $2n\pi/Lx$, the free energy changes by

$$dF = \frac{dF}{dk}\frac{2\pi}{Lx}Lx\,dy \ .$$

This change of energy corresponds to the work of the Peach-Köhler force $f_{\mathrm{PK}} = dF/dy = 2\pi dF/dk$ acting on the dislocation. The Peach-Köhler force governs the motion of the defect, but the speed is limited by friction processes. For constant speed we have $f_{\mathrm{PK}} = f_\nu$. The damping force f_ν depends on diffusion in the nonequilibrium medium [4.40].

4.3 Self-Structures

The origin of this term is related to its pictorial oscillatory analogue. The oscillations are often classified as free, forced and self-oscillations. We shall apply this classification to structures. Then, free structures are things like the ring vortices in ideal liquid flows, rings of smoke, etc. Examples of forced structures are, in particular, the ring rolls that repeat the shape of the cylindrical container in which the Rayleigh-Bénard convection is observed [4.5]. Self-structures are less trivial patterns, therefore we must first define them exactly. Self-structures are localized spatial formations that are stable in dissipative nonequilibrium media and do not depend (within finite limits) on boundary or initial conditions. We are already acquainted with self-oscillations which may have a family of conservative (free) oscillations as generating solutions; small dissipation and energy supply introduced into such a system merely "choose" a definite motion but do not change its shape appreciably. In a similar way self-structures in weakly nonequlibrium media may inherit the properties of free structures. Well-known examples of such "quasi-conservative" structures are the solitons that live due to the nonequilibrium of the medium like Rossby solitons, which include apparently the Big Red Spot of Jupiter, see Sect. 2.1 and [2.37].

Technical knowledge it not enough. One must transcend techniques so that the art becomes an artless art, growing out of the unconcious.

Daitsetsu Suzuki

4.3.1 Convective Self-Structures

We shall first demonstrate self-birth and stable existence of self-structures, using as an example experiments on thermocapillary convection. The experiments were essentially as follows [4.41]: the dynamics of convective structures was investigated in thin horizontal layers of silicon oil heated inhomogeneously from below. A cylindrical container, with copper heating elements of different shapes incorporated into its bottom was used. The temperature of the heater was maintained at a constant level by means of a thermostat, the upper boundary was free. The shape of the container boundaries was varied by means of additional partitions. The emerging structures were visualized by a thin aluminium powder in the liquid.

The experiments demonstrate the existence of solitary self-structures in the form of hexahedral cells, whose parameters are (within the given degree of nonequilibrium) independent of the shapes of container boundaries and of the specified temperature inhomogeneity.

Figure 4.14 displays the thermocapillary structures, observed in a cylindrical container (for different layer thicknesses d and a constant temperature difference δT). For a small depth with $d \ll L$ (L being the diameter of the heater) we see that the cells fill the entire region occupied by the heater; as d increases they grow in size and for finite variations of d a solitary cell structure is established, independent of the initial conditions. With a further increase of d, this regime is replaced by another one of large-scale convection in the form of several vortices,

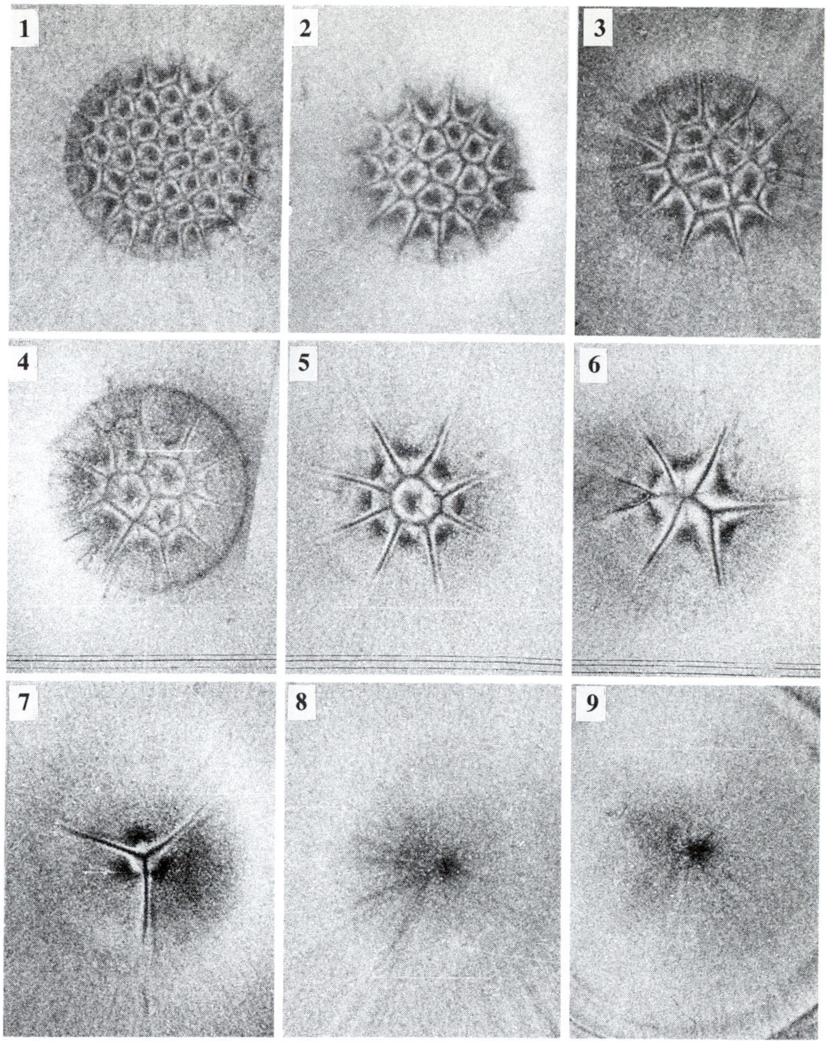

Fig. 4.14. Different kinds of structures for different depths of the layer. The pictures show the effect of varying depth of the liquid layer (which increases with the labels given on the figures)

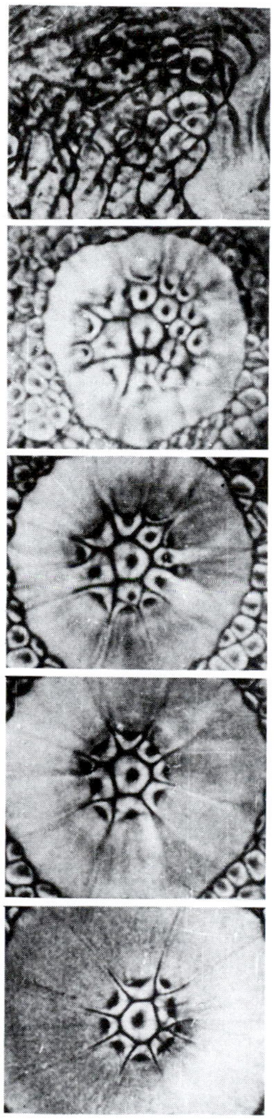

Fig. 4.15. Formation of a convective octahedron in an initially mixed fluid (the time t as measured in minutes increases from the *top* to the *bottom*)

having sizes comparable to the one of the container, the shape of its boundaries determines the one of the cylindrical waves, see Fig. 4.14.

The dynamics of the birth of a solitary cell is shown in Fig. 4.15. The parameters of the self-structure are seen to be independent of the initial conditions. For comparison, the dynamics of structures arising under slightly modified conditions was investigated, the only deviation being the use of a square-shaped heater. Experiment indicates that the resulting structure remains practically unchanged.

The observed self-structures arise for the variation of the medium parameters within finite limits and demonstrate for a certain increase of supercriticality nontrivial dynamics. In particular, we have observed a nearly periodic regime of the interaction of two cells, in which they die and are reborn alternatively, retaining their original shape, see Fig. 4.16.

On the basis of these experiments, we can state that the observed self-structures are the eigenmodes of a nonlinear nonequilibrium medium and are not related to the presence of boundaries.

4.3.2 Localization Mechanisms

Our examples show that self-structures of universal shape can, actually, be established in nonequilibrium media. However, the localization of the field structure is in most experiments determined by the inhomogeneity of the medium. Is the self-generation of localized structures, independent of organizing forces, also possible in isotropic and homogeneous (!) media? We can prove that the self-generation of such structures is possible, if we can construct a phenomenological model, within which localized structures exist.

Using (4.7) we shall now construct such a model describing the formation of ensembles of structures and the motion of defects in two-dimensional models of three-dimensional convection. Unlike (4.7), we shall in this model account for the coordinate dependence of μ:

$$\partial_t u = \left[\mu(\boldsymbol{r}) - \left(q^2 + \nabla^2\right)^2\right] u + \beta u^2 - u^3 . \qquad (4.17)$$

Computer experiments show that this equation may serve as a model of the birth of polyhedrons in a fluid layer with localized heating, for $\beta \neq 0$ and the corresponding structure of the increment $\mu(\boldsymbol{r}) = -\alpha + v(r)$. The calculated shape of the cells agrees well with the pattern observed experimentally, see Fig. 4.15.

In order to construct a self-consistent model for the description of the birth of localized self-structures in an isotropic medium (i.e. a model without artificially produced inhomogeneities), (4.17) must be supplemented with an equation for $v(r,t)$ with coordinate independent parameters. This extended equation should take into account the connection with u and support spatially localized solutions corresponding to inhomogeneous heating. Thus it is natural to employ the equation for a homogeneous nonlinear medium with heat release, diffusion and an additional source:

$$\varepsilon \partial_t v = v - \gamma v^3 + D \nabla^2 v + \delta u . \qquad (4.18)$$

For $\delta = 0$ this equation supports localized solutions. However, they are nonstationary: depending on the type of perturbations, excitations either collapse or are transformed into spatially homogeneous ones $v^0 = \gamma^{-1/2}$. Nevertheless, in a self-consistent problem such localized solutions may be stationary and stable, for example, when the heat release at the periphery of the localized solution $v(\boldsymbol{r}, t)$ is suppressed by the field $u(r,t)$. Thus, we arrive at the following model system [4.42, 43] with $q^2 = 1$

$$\begin{aligned} \partial_t u &= \left[(v - \alpha) - \left(1 + \nabla^2\right)^2\right] u + \beta u^2 - u^3 , \\ \varepsilon \partial_t v &= v - \gamma v^3 + \delta u + D \nabla^2 v , \quad \varepsilon \ll 1 . \end{aligned} \qquad (4.19)$$

To verify our suppositions on the self-generation of localized patterns in a homogeneous medium, we shall first consider the computer results for small β. In this case one observes actually such self-structures – they take the shape of disks with their characteristic sizes and sta-

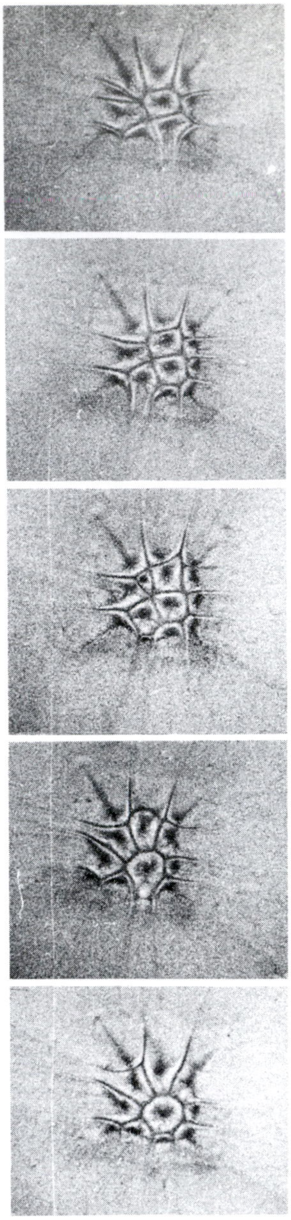

Fig. 4.16. Periodic self-excited oscillations in a small ensemble of convective cells

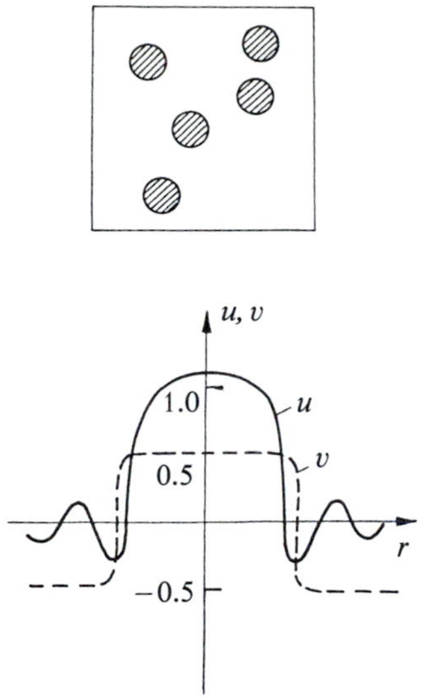

Fig. 4.17. Localized disc-like structures in model (4.19)

tionary intensities determined only by the parameters of the medium and not depending on the initial or boundary conditions in the region considered, see Fig. 4.17.

Number and mutual location of the self-structures, emerging in a sufficiently extended two-dimensional medium described by (4.19), are determined by random initial conditions. However, they cannot converge by more than the suppression scale l, that corresponds to the characteristic size of the region bordering on the self-structure, in which the field u is negative, Fig. 4.17. The existence of this region is determined by the peculiarities of the diffusion, the terms $\nabla^2 u$ and $\nabla^4 u$ in (4.17). In the suppression region the nonequilibrium medium is not excited, which ensures the stability of the self-localization effect of the two-dimensional fields u, v.

The nontrivial shape of the self-structures is explained by the diversity of the linear excitations that serve as sources for the further nonlinear growth and formation of self-structures. Regular polyhedrons are the simplest nontrivial structures with a symmetry centre. They emerge as a result of the interaction of two modes – radial and azimuthal – of a circular membrane. In particular, self-structures in the form of solitary rectangles and hexahedrons as discussed above (see Fig. 4.18), can be considered as the result of the interaction of these modes. According to experiments, the nonlinearity, related to the temperature dependence of surface tension, plays a dominating role in the generation of solitary polyhedrons. This nonlinearity is taken into account by the term proportional to βu^2 in model (4.19). Such a nonlinearity ensures the simultaneous generation of modes which compete with each other for small β.

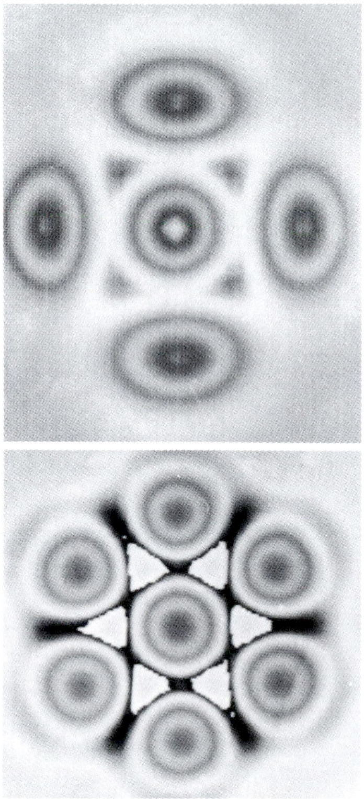

Fig. 4.18. Secluded rectangle and hexahedron in the system described by (4.19) with $\beta \geq 1$

4.3.3 Self-Structures in Three-Dimensional Media

The particle-like solutions for nonlinear fields have been intensively sought for quite a long time, nevertheless, at present there are only a few examples of stable localized solutions. The discovery of solitons has greatly contributed to the advance in this field. However, static stable solitons are only supported by one-dimensional systems. Long-lived two- and three-dimensional soliton(-like) or solitary solutions are as a rule nonstationary [4.44, 45]. The three-dimensional solitary static solutions can be expected to be stable within the nonlinear field equations describing processes in nonequilibrium dissipative media. It is in these media that the self-generated stable spatial lattices are observed, lattices with square cells, like honey-comb patterns, etc. The nonequilibrium medium described by (4.19) supports two-dimensional localized structures. It appears natural to seek for three-dimensional "particles" within the same equations (assuming $\nabla^2 = \partial_{xx} + \partial_{yy} + \partial_{zz}$). Since the fields, described by the Swift-Hohenberg type potential, see for example (4.16), are multistable, we can expect that different localized structures will arise for identical parameters but different initial conditions within the three-dimensional model (4.19) or in

$$\partial_t u = -u + \beta u^2 - u^3 - \left(1 + \nabla^2\right)^2 u \ . \tag{4.20}$$

Recent computer experiments [4.46, 47] have shown that this model describes actually the self-generation and stable existence of three-dimensional particle-like solutions of different topologies. Three "elementary particles" have been found – a sphere, a torus, and a spherical lattice (a "baseball"), having a characteristic size ~ 1, see Fig. 4.19. Stable solutions in the form of the bound states of elementary particles (equal or different ones) are observed under appropriate initial conditions, see Figs. 4.20. The elementary particles have arbitrary spatial orientations depending only on the initial conditions. Their topology and size are universal and do not change with varying boundary conditions and size of the region. Note that the nonlinear field (4.19) or (4.20), in a definite region of initial conditions, allows for the formation

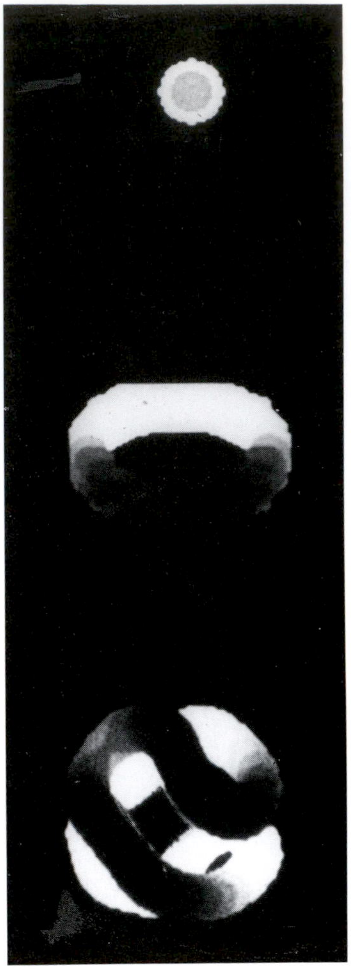

Fig. 4.19. "Elementary particles" of model (4.19): sphere, torus, "baseball" (with a pattern similar to the one of a tennis ball)

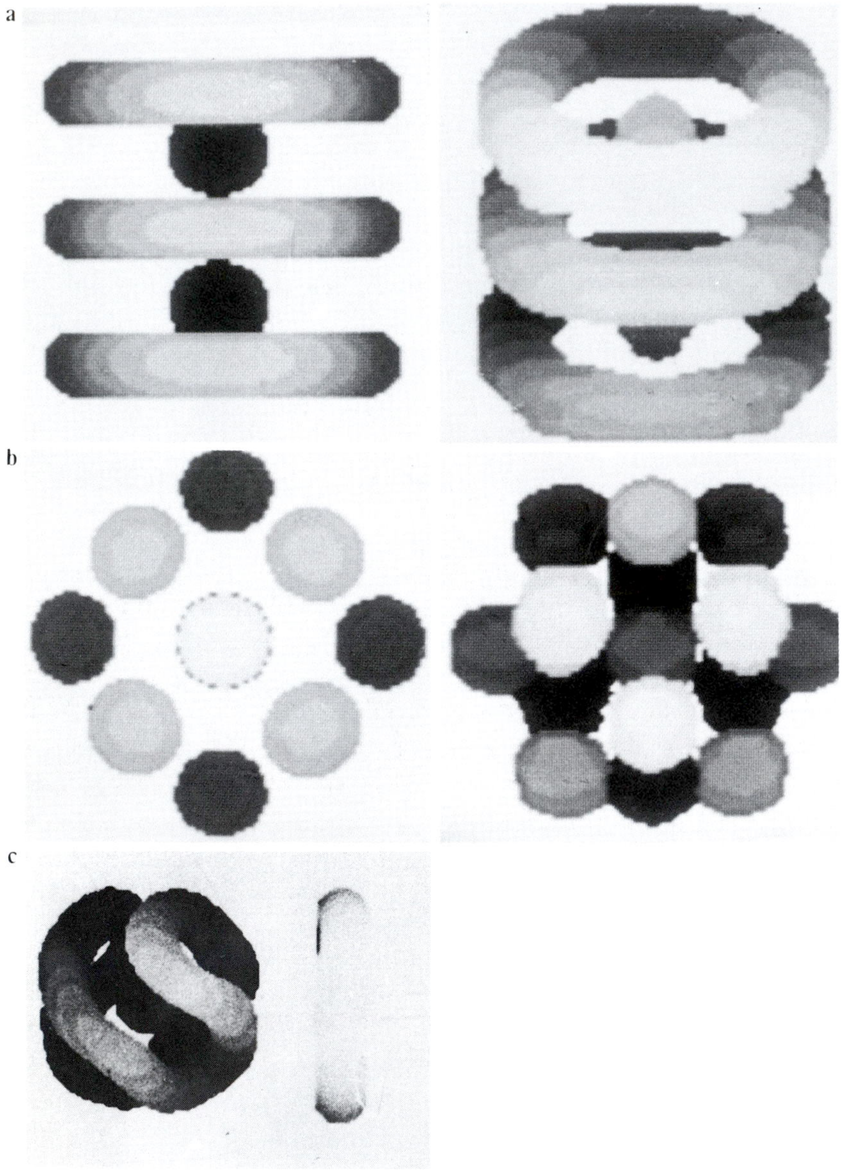

Fig. 4.20a–c. Bound states of "elementary particles" ($\beta \approx 1.5$):
a) three tori and two spheres;
b) clusters of spheres;
c) torus, "baseball" and sphere

Fig. 4.21. Metastable structure in the form of a spiral

of various structures that are not immediate bound states of "elementary particles", for example, the spiral shown in Fig. 4.21. However, such structures are not attractors (in our case trivial equilibrium states) of the system under consideration and go over into bound states of elementary particles as $t \to \infty$.

We would like to emphasize that the physical mechanisms that lie at the foundation of the self-generation of the particle-like solutions of three-dimensional fields (and that are related to the interaction of components and spatial dispersion) are, in fact, quite general ones and can be realized in various nonequilibrium media.

4.3.4 Interaction of "Elementary Particles"

In a simplified picture various experiments will lead us to the conclusions that the interactions of localized particle-like structures can be classified into "weak" and "strong" ones. In the first case, the localized structures including solitons are rather far away from each other so that the field of one structure or particle at the centre of the other one can be considered to be weak. We use this fact in the construction of a theory for adiabatic interactions, see below. Strong interactions are interactions in which the particle-like solutions change qualitatively and/or are born and die when colliding with one another. The result of strong interactions depends on the particle's prehistory and on the field properties. We believe, that we can exclude strong interactions from the description of nonlinear dynamics of ensembles of structures, and, instead, use certain rules according to which they are mutually transformed and enter a subsequent stage of weak interactions, that can be described approximately by taking into account the conservation of the structure of the localized object.

Obviously, not all nonlinear dynamics of localized structures adhere to the interaction scheme described above (in which "adiabatic" corresponds to weak and "bifurcational" to strong), however experiments yield many suitable examples. In this connection, we find it important to formulate sufficiently general models, that would describe experimental situations, on the one hand, and allow for a more or less complete analytic and computer investigations, on the other hand. As such models we will use the generalized gradient systems of the form [4.47]

$$\hat{L}(\partial_t)u = -\frac{\delta F}{\delta u} + \varepsilon\Phi(\boldsymbol{r}, t, u) \ , \quad \varepsilon \ll 1 \ . \tag{4.21}$$

Here $u(\boldsymbol{r}, t)$ are physical variables; F is a functional having the meaning of the free energy of the system (medium, field) and $\Phi(\boldsymbol{r}, t, u)$ is the nonlinear operator taking into account the action of external fields, the deviation of (4.21) from a potential system, etc.

For $\hat{L} = \partial_t$ and $\varepsilon = 0$ the dynamics of the particle-like solutions of (4.21) is determined by $\partial_t u = -\delta F/\delta u$. This is a traditional gradient system all solutions of which are known to be static ones as $t \to \infty$.

The local minima of the functional F correspond to these static solutions. It follows directly that in such media localized structures either come to a stop or go away to infinity or "die", say, when merging with each other. However, the transition to this final stage may be a diverse and multistep process. In this connection, we would like to draw your attention to the computer experiments on the two-dimensional model (4.10, 16), in which the dynamics of the defects – localized structures – was investigated against the background of a periodic lattice. Figure 4.8 illustrates the $F(t)$ dependence and shows that the transition to the state with $F = F_{\min}$ is, actually, a sequence of adiabatic steps separated by fast motions corresponding to the merging of defects.

The existence of stationary localized structures evidently does not depend on the form of the operator $\hat{L}(\partial_t)$ (they are solutions to (4.21) for $\delta F/\delta u = 0$, $\varepsilon = 0$), therefore it is not surprising that conservative media ("Hamiltonian" fields) and nonequilibrium dissipative media can support the same particle-like solutions. However, for the Hamiltonian those of them that do not exactly correspond to the minimum of F will not tend to static solutions, but oscillate, possibly even chaotically, like a marble rolling without friction over the bottom of the valley.

For small perturbations ($\varepsilon \ll 1$), the particle-like structures will generally no longer be static; they will interact with external fields, walk randomly, deform slowly, etc.

In every department of physical science there is only so much science, properly so called, as there is mathematics.

Immanuel Kant

As an example of weak interactions we shall consider the dynamics of bound "elementary particles" of the sphere type in model (4.19, 20), where $\nabla^2 = \partial_{xx} + \partial_{yy} + \partial_{zz}$. The asymptotic method can be used to derive for the coordinates of the centres of the spheres $\boldsymbol{r}_{0j} = (x_{0j}, y_{0j}, z_{0j})$ ordinary differential equations of the form [4.47]

$$\frac{d\boldsymbol{r}_{0j}}{d\tau} = \nabla_{\boldsymbol{r}_{0j}} \sum_{l \neq j} \operatorname{Re} \left\{ \frac{\exp\left[ik|\boldsymbol{r}_{0j} - \boldsymbol{r}_{0l}|\right]}{|\boldsymbol{r}_{0j} - \boldsymbol{r}_{0l}|} \right\} . \tag{4.22}$$

If we have only two spheres then they will move along the line connecting their centres, until a stable equilibrium state – a bound state – is established. There exists an infinite number of stable bound states even for two "particles", the more so for several of them. They may have the form of rectilinear polyhedrons of spheres, see Fig. 4.20b, of periodic and "quasi-crystalline" lattices, and so on.

The dynamics and interactions of "elementary particles"in the conservative system

$$\partial_{tt} u = -u + \beta u^2 - u^3 - \left(1 + \nabla^2\right)^2 u \tag{4.23}$$

whose stable static solutions coincide with those considered above are much more diverse. The particles may, in particular, rotate relative to one another forming planet-like systems, chaotically approach and leave each other, etc.

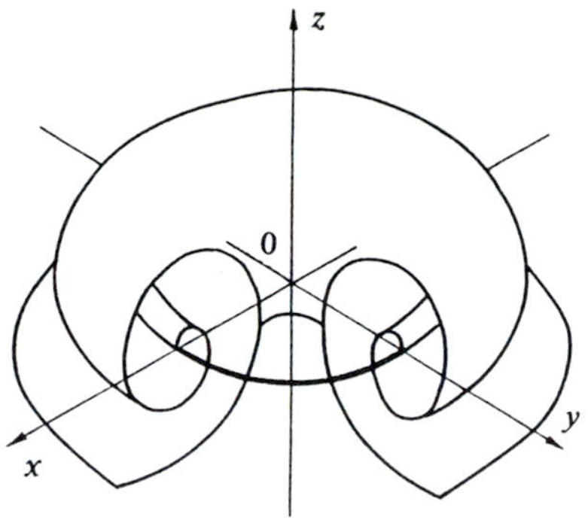

Fig. 4.22. Spiral ring

4.3.5 Birth and Interaction of Spiral Waves

In Fig. 4.19 we have seen for the case of torus and "baseball" that anisotropic structures may appear in homogeneous isotropic media. Rotating spiral vortices are the most pictorial example of such anisotropy. These structures are so frequently encountered in different experiments, that they might be considered to be quite elementary and universal and not connected to the physical origin of a nonequilibrium medium, see Fig. 4.22.[6] Why are the spiral vortices so widely met and what mechanisms are responsible for their formation? The universality of spiral patterns can be shown [4.49] to be related to quite general topological properties of one-parametric families of functions $H_t(x, y)$ = const that describe the spatial images (snapshots) of the field in two-dimensional nonequilibrium media. We will now consider briefly the mechanism of the birth of spirals and their transformation into concentric waves. As an instructive example we take media described by the "Λ-ω model" or by a two-dimensional Ginzburg-Landau equation. We will see that the spirals arise as the degree of instability in the medium is increased. This results from the hierarchy of spatial instabilities with different types of symmetry.[7]

Structures of a two-dimensional field that allow for direct observation (see, e.g. [4.50]) are usually made visible by contrast or by colour. In particular, the concentration structures observed in a two-dimensional reactor, where an autocatalytic chemical reaction takes place, are colour patterns [4.51]. For the sake of definiteness, we shall restrict our consideration to such structures by assuming that the kinetics of the reaction is determined by the interaction of only two components:

$$\begin{aligned} \partial_t n_1 &= n_1 - \left(n_1^2 + n_2^2\right)(n_1 - \beta n_2) + \kappa\left(\nabla^2 n_1 + C\nabla^2 n_2\right) , \\ \partial_t n_2 &= n_2 - \left(n_1^2 + n_2^2\right)(n_2 + \beta n_1) + \kappa\left(\nabla^2 n_2 - C\nabla^2 n_1\right) , \end{aligned} \tag{4.24}$$

where $n_1(x, y, t)$ and $n_2(x, y, t)$ are the densities of these components and κ describes the strength of the diffusion. System (4.24) is a variety of the well-known Λ-ω system used for modelling the nonlinear dynamics of various nonequilibrium media [4.29]. Substituting $u = n_1 + in_2$ into (4.24) we obtain a two-dimensional Ginzburg-Landau equation.

The colour distribution in concentric structures will be determined by superimposing the colours of the individual components. Since the topology of structures (the position of the lines of constant colour) must be invariant with regard to proportional variations of the intensity of the individual colours, the problem of investigating the structures reduces to the analysis of the level lines of the function

6 Experiments on the chemical Belousov-Zhabotinskii reaction and in some biological media [4.48] demonstrate three-dimensional spirals or vortex rings like in Fig. 4.22.

7 Here, again, we will employ the evolutionary or "embryologic" approach proposed by A.A. Andronov, but for the investigation of self-structures and not self-oscillations.

$$\frac{n_1(x,y,t)}{n_2(x,y,t)} = H_t(x,y) = \text{const} . \qquad (4.25)$$

We will be interested in those field configurations $H_t(x,y)$ that cannot be transformed into spatially homogeneous ones just by using continuous deformations. At singular points the values of fields are not defined. The described configurations are determined by the presence of such singular points in $H_t(x,y)$, of "separatrices", etc. which correspond to elementary structures in the form of spirals, vortices, spines and the like.

Bifurcations – transformations of one structure into another one – correspond to a change of the topology of the level lines of $H_t(x,y)$ as determined by (4.24) due to a variation of a parameter (or time). We will find these changes by varying the diffusion parameter κ. The function $H_t(x,y)$ contains a lot of important information for us and is related to the phase φ of the complex field $u = |u|e^{i\varphi} = n_1 + in_2$ by the relation $H_t(x,y) = \arctan \varphi(x,y,t)$. We shall investigate the field structure determined by (4.24) with periodic boundary conditions and assume

$$u(x,y) = u(x+L,y) = u(x,y+L)$$

to hold.

In addition to that we shall take into account the fact that nontrivial solutions of (4.24) are possible only for $\beta C > 1$. It may be verified [4.29] that in the considered medium for the values $\kappa > \kappa_0 = (\beta C - 1)L^2/\pi^2(1+C^2)$ of the diffusion and for $\beta C > 1$ only a spatially homogeneous regime of oscillations with $u(x,y,t) = \exp[-i\beta t]$ can be established.

Fig. 4.23. Periodic lattice of vortices and its transformation into a spiral lattice (in (x,y)-space)

For $\kappa < \kappa_0$ the regime loses its stability and against its background four waves of the form $\exp[\pm ik_0x]$ and $\exp[\pm ik_0y]$ build-up. The function $H_t(x,y)$ has no singularities for such field distributions. $H_t(x,y)$ describes the simplest regular spatial structure in the form of a periodic lattice of "vortices" (see Fig. 4.23) that is stable when $\kappa \lesssim \kappa_0$.

As the parameter κ decreases still further, four more waves having wave vectors located at an angle of 45° with regard to the initial ones are excited against the background of the lattice. For sufficiently large amplitudes of the waves, the function $H_t(x,y)$ has singularities, i.e. new elements of the structure will emerge as κ is reduced. The form of the new elements is determined by the behaviour of the level lines of the scalar field $H_t(x,y)$ in the immediate vicinity of the singularities. The general topological properties of the configurations of interest are studied by the geometrical theory of foliations [4.52]. In particular, due to the analyticity of the function $n_{1,2}(x,y,t)$ it follows from this theory that in our case (by virtue of index conservation) the most typical bifurcation is the transformation of a pair of "vortices" into a pair of nodes (or spirals) as κ varies, see Fig. 4.24. Such a bifurcation of structures is also possible when the solution changes in time (when $\kappa < \kappa_0$). In this case, the evolution leads to vortices that transform into

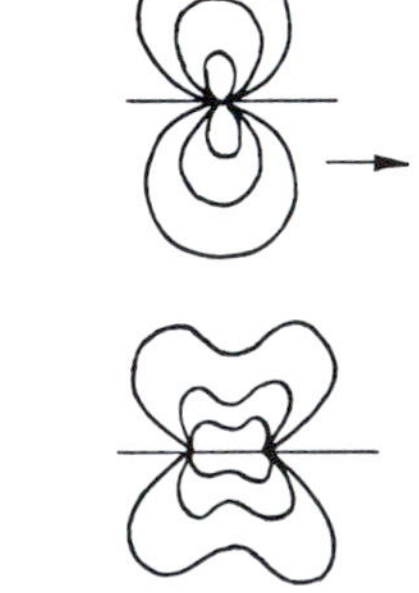

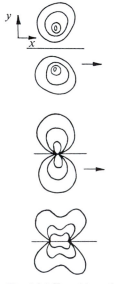

Fig. 4.24. Transition of a pair of vortices into a pair of nodes (spirals)

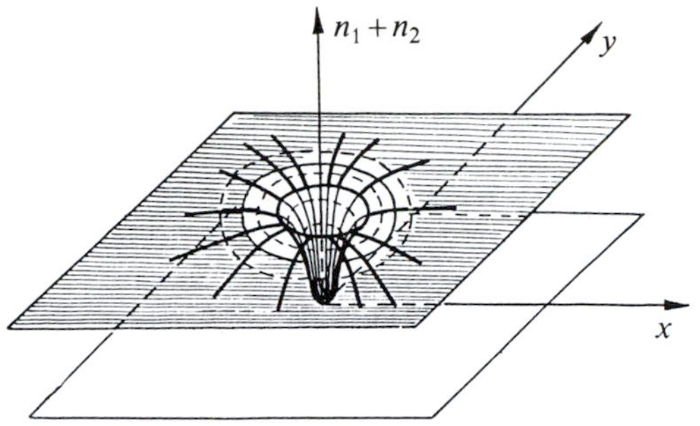

Fig. 4.25. Intensity distribution in the vicinity of the center of a spiral

spirals. In fact, this is an example of a "strong" interaction between structures.

We would like to note that the zeros of the intensities at n_1 and n_2 correspond to the centres of the spirals. The intensity distribution near the centre of the spiral is very similar to a dark soliton, see Fig. 4.25. When there are just a few spirals (dark solitons), we can give the equations of motion for their centres [4.53]. These equations are similar to the ones obtained for the interaction of heavy particles, see (4.22).

4.4 Attractors – Memory – Learning

4.4.1 How to Remember

Learn to be careful with your judgement:

That is the biggest fool thing we have ever done. The [atomic] bomb will never go off, and I speak as an expert in explosives.

Adm. W. Leahy
to President Truman 1945
from R.L. Weber
"A Random Walk in Science"
(Institute of Physics, London, Bristol 1977) p.67

We have already mentioned that the realization of one or the other static state of a multistable nonequilibrium medium depends on the initial conditions. This implies that different continua of initial conditions give rise to different static patterns. In the corresponding phase space these different states are mapped by attractors (equilibrium states for static patterns). The boundaries of the attracting regions of the attractors determine possible variations of the initial conditions under which the medium "demonstrates" the same picture for the limit $t \to \infty$. If we want to observe a different spatial picture, we must go over to the region of attraction of another attractor. This can be accomplished by means of an external field, that contains the needed information (not all of it will necessarily be useful!) on the desired future state. Since we only have to cross the region of attraction of the needed attractor, the magnitude of action is in this case less significant than its spatial structure, i.e. it is the informative rather than the force aspect of the action which is dominant here.

The thoughtful reader must have not only surmised that Nature itself has destined nonequilibrium multistable media for remembering and identifying information, but must have already started thinking about the ways for putting this almost apparent idea into practice. Over the last decade, much has already been done in this field of nonlinear physics. In particular, there are examples of the application of natural and man-made nonequilibrium media in systems developed for remembering, learning and identifying images.

We believe that the developments connected with the application of some artificial media known as spin glasses [4.54] are the most promising. These "glasses" are dynamical systems of bounded lattices of "spins" each of which may be in two states. The bonds between the spins are formed in the process of learning such that an attractor in the phase space of the system should correspond to each image to be remembered, i. e. lattices of bound spins are formed. The more extended the region of attraction of such an attractor, the more reliable is the process of the identification and reproduction of the image by one of its parts or in the presence of distortions. The transition of the considered dynamic system from the region of attraction of one attractor to the one of another, induced by the image of the "pretender", corresponds to the transition from the reproduction of one image to the reproduction of another one. Of course, not every proposed image may find an appropriate attractor in the available dynamic system – a spin lattice with specified bonds. Therefore, the natural step to follow in the development of these ideas is the "predetermined" formation of bonds between individual spins using the needed image. Very much like the work of the brain!

All things are difficult before they are easy.

It is very likely, that advances in modern computer technology will make it quite feasible to investigate real phenomena not through an intermediate stage of their idealization in terms of infinite and continuous mathematics, but to go immediately over to discrete models. This holds in particular for the investigation of systems having complicated organization and being able to process information. Exploiting his brain as the reality allotted to him by God, a mathematician could take no interest in the combinatory foundations of its working. However, the artificial intellect of machines must be created by Man and he will inevitably have to go deep into combinatory mathematics. These words of *Andrei N. Kolmogorov*, an outstanding mathematician of our time, stated a quarter of a century ago [4.55], are still valid. We may say that the understanding of the relationship between continuous and discrete, as applied to the working of the brain, is spiral-like. We are now at the "return winding" of the spiral – there is a tendency to consider the work of neuron ensembles of 10^9 elements (each being connected with $10^2 - 10^4$ others) as the dynamics of a multicomponent nonequilibrium medium. Let us assume that one of the fields in this medium changes much more slowly than the others. If we then subject it to the action of an external field (image) while it is in the process of learning, we will be able to form the free energy functionals F of the other fields such that each picture to be remembered could give rise to its own local minimum of F.

In this medium the number of different pictures it will memorize and identify corresponds to the number of resolvable minima of F (at a given level of fluctuations) we will be able to create. A simple example for such a medium which can easily be arranged is a conventional TV-set with a video camera.

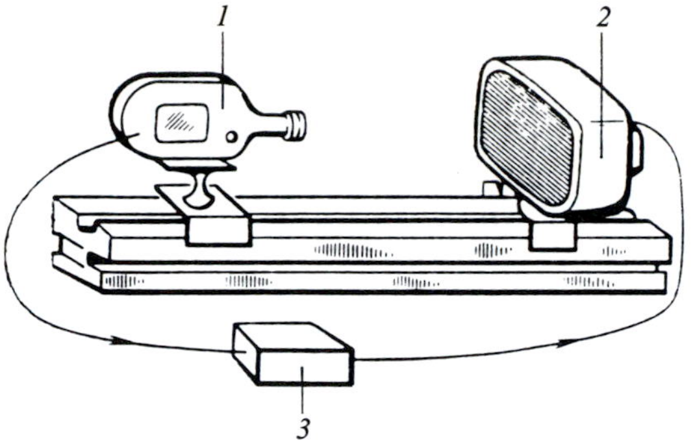

Fig. 4.26. Schematic illustration of the TV-analogue "camera + TV + feedback"; (*1* – telecamera, *2* – monitor, *3* – amplifier of the feedback system)

4.4.2 "Camera + TV + Feedback" Analogue

What happens if a video camera connected with a monitor (see Fig. 4.26) faces the screen of this monitor? Evidently, we will see on the screen the screen of the monitor, inside it, there will be another screen of the monitor, and still another and another, until our eye or apparatus will be able to resolve details. Now we will increase the amplification and bring the video camera so close to the screen that the monitor seen on the screen will have the size of the screen. In this case no initial information will remain on the screen. It will show something of its own! This video system gives birth to images that are at times quite unexpected and attract the attention of the observer (see Plates 43 to 52 of Chap. 6 *Nonlinear Physics*) – the structures on the screen are continuously moving, emerging and fading away. They seem to be real living beings.

Now we shall attempt to look into the mechanisms responsible for this video pattern. Let us first recall the well-known situation in which a microphone connected via an amplifier to a loud-speaker is installed in a lecture- or concert-hall. Careless handling of amplifier and/or microphone leads to self-excitations or self-oscillations of the system. In such a situations the system forgets completely about its initial acoustic information and quite independent stable acoustic signals are produced. Something similar occurs in our "TV + camera + feedback" system, only in this case we go over from the realm of acoustic signals to the one of video images – this is a generator of pictures!

In such a system the TV screen is a model of a two-dimensional nonequilibrium medium, which lends itself for investigating various self-organization phenomena and for modelling the memory and the process of learning [4.56–59]. The parameters of this nonequilibrium medium may be varied within broad ranges, simply by using the "contrast" and "brightness" controls on the monitor and the "focus" and "distance" (translocator) on the camera. This medium may be both softly self-excited – when an arbitrarily weak initial optical action is able to ignite the screen and, hence, to create a stationary or continuously evolving (living) image – and a threshold system – when a light source of finite intensity is needed to initiate an excitation in the medium (in chemistry and biology such media are often referred to as excitable media).

The Moving Finger writes; and, having writ,
Moves on; nor all thy Piety nor Wit
Shall lure it back to cancel half a Line,
Nor all thy Tears wash out a Word of it.

Rubáiyát of Omar Khayyàm 1859
English by E. Fitzgerald

The arrangement depicted in Fig. 4.26 may have various applications. One of them is the solution of specific problems, for instance, the calculation of the velocity of a propagating excitation (e.g. flames) in media with defects (inhomogeneities).

We will now consider the potentialities of a black-and-white variety of the TV-analogue in terms of the analysis of the behaviour of structures in two-dimensional nonequilibrium media with an N-shaped nonlinear characteristic of the element of the medium, see below. The equations, modelling such a single-component system, may describe: the motion of the gas phase in the film regime of boiling in heat release elements [4.60], the dynamics of ensembles of dividing cells in living tissues [4.61], various oxidation reactions in plane catalizers [4.62], etc. The proposed model proved to be adequate, in particular, for well-studied self-wave processes like the propagation of flames moving around an obstacle which were simulated with the TV-analogue. Using this model, we are able to study the mechanism of the interaction of localized structures, in particular, related to the transfer of excessive excitation, as well as the processes of the emergence of the elementary structures and quasi-one-dimensional ensembles as formed by these structures, and so on. The basic equations for the description of the TV-analogue are difference equations, averaged over the inertia or response time $\tau_{\rm targ}$ of the camera target, that relate the potential u_n on the surface of the camera target at the point (x, y) to the image brightness I_n on the TV screen at (x, y), at the times $t_n = t_0 + n\tau_c$ (τ_c is the time needed for the formation of the picture and $\tau_{\rm targ} \sim 10\tau_c$):

$$\begin{aligned} u_{n+1}(\boldsymbol{x}) &= \beta \hat{R}_{\rm targ} u_n(\boldsymbol{x}) + f_1[\hat{R}_0(I_n(\boldsymbol{x}))] \ , \\ I_{n+1}(\boldsymbol{x}) &= f_2[u_{n+1}(\boldsymbol{x})] \ , \quad \beta = 1 - \tau_c/\tau_{\rm targ} \ . \end{aligned} \tag{4.26}$$

Here

$$\hat{R}_0(I_n(\boldsymbol{x})) = \int_{S_{\rm s}} R_0(\boldsymbol{x} + \boldsymbol{\zeta} - \boldsymbol{\xi}) I_n(\boldsymbol{\xi}) d\boldsymbol{\xi} \ ,$$

$$\hat{R}_{\rm targ}(u_n(\boldsymbol{x})) = \int_{S_{\rm targ}} R_{\rm targ}(\boldsymbol{x} - \boldsymbol{\xi}) u_n(\boldsymbol{\xi}) d\boldsymbol{\xi} \ ,$$

hold and $S_{\rm s}$ and $S_{\rm targ}$ are the planes of TV screen and camera target, respectively. The function f_1 describes the transformation of the optical image to the electronic one, and f_2, the electron–optical transformation. The function R_0 determines the blurring of the image in the optical channel of the system and has approximately a Gaussian form, similarly to the function $R_{\rm targ}$ that characterizes the spreading of the charge on the target; the parameter $\boldsymbol{\zeta}$ is related to the relative shift between the optical axes of camera and TV-set.

Bearing in mind that we want to model processes with characteristic times much greater than τ_c, the averaging over the interval $t \sim l\tau_c$ ($l \gg 10$) leads us to continuous time

$$\tau_c \partial_t u = -(1-\beta) \int\limits_{S_{\text{targ}}} u(\boldsymbol{\xi}) R_{\text{targ}}(\boldsymbol{x}-\boldsymbol{\xi}) d\boldsymbol{\xi}$$
$$+ f_1 \left[\alpha(\boldsymbol{x}) \int\limits_{S_s} f_2(u(\boldsymbol{\xi})) R_0(\boldsymbol{x}+\boldsymbol{\zeta}-\boldsymbol{\xi}) d\boldsymbol{\xi} \right] . \tag{4.27}$$

The inhomogeneity $\alpha(\boldsymbol{x})$ of the parameters of the modelled medium may be produced by means of masks placed on the plane conjugate to the one of the electron image. Then, disclosing the form of the operators and taking into account the linearity of the active section $f_1(u)$, i.e. $f_1(u) = Ku$, we obtain the nonlinear equation with diffusion

$$\partial_t u = \mathcal{F}(u, \boldsymbol{x}) + \boldsymbol{J}(u, \boldsymbol{x}) \nabla u$$
$$+ \alpha(\boldsymbol{x}) \left[\delta^2 \nabla (D(u) \nabla u) + \boldsymbol{\zeta} \nabla (D(u) \boldsymbol{\zeta} \nabla u) \right] ,$$
$$\delta^2 = 0.5 \int\limits_{S_s} \boldsymbol{\xi}^2 R_c(\boldsymbol{\xi}) d\boldsymbol{\xi} ,$$

where

$$\mathcal{F}(u, \boldsymbol{x}) = -\left(\frac{1}{\tau_c} - \frac{1}{\tau_{\text{targ}}} \right) u + \frac{K\alpha(\boldsymbol{x})}{\tau_c} f_2(u) ,$$
$$\boldsymbol{J}(u, \boldsymbol{x}) = \boldsymbol{\zeta} \frac{K\alpha(\boldsymbol{x})}{\tau_c} f_2''(u) , \quad D(u) = \frac{K f_2'(u)}{\tau_c} .$$

It has a more general form than (4.4–7) which is usually employed for the description of various processes of birth, propagation and interaction of structures and other excitations in nonequilibrium media.

The function $\mathcal{F}(u, \boldsymbol{x})$ constructed by means of direct measurements, is an N-shaped curve, whose slope and average value are determined by the "contrast" and "brightness" controls, and $f_2'(u)$ is a trapezoidal function.

Flow with whatever may happen and let your mind be free: Stay centered by accepting whatever you are doing. This is the ultimate.

Chuang Tzu

It is known [4.63] that stationary "transition" fronts (from one equilibrium state to another) may propagate in systems of the form (4.27). Such a transition (or switching) for one-dimensional waves (see Chap. 1 and Fig. 1.1a) is described by the saddle separatrix in the phase plane of the equation of a nonlinear oscillator with friction. In our case, this is the transition from the homogeneously illuminated TV screen to a dark one or vice versa. Note that experiments with such a TV-analogue are usually performed without external illumination sources; they may only be needed for setting definite initial conditions ("initiation") at the required points of the TV screen. The "transition" waves that were obtained on the TV-analogue, including those observed for flames moving around various obstacles (masks on the TV screen), agree well with the solutions found in numerical modelling.

When modelling standing self-structures, the shape of the function should be chosen by means of the "brightness" and "contrast" controls such that the velocity of the transition fronts from the "black" to the "white" (or vice versa) is nearly zero. The self-structures will then be localized light cells.

Let the initial state be a screen with a dark and an illuminated half. As the parameters of the "medium" approach the values of interest, the homogeneous interface loses stability and is modulated with a characteristic scale l_0. The modulation depth will subsequently increase and cells will emerge. When the active part of the screen is a stretched rectangle, the cells arrange themselves into a homogeneous chain.

Actually, the parameters of this chain depend neither on the initial nor on the boundary conditions. For example, if we "remove" one or several intermediate cells, then the remaining cells will move closer to each other until they reproduce the initial step (the number of elements in the chain will now be smaller), Fig. 4.27.[8]

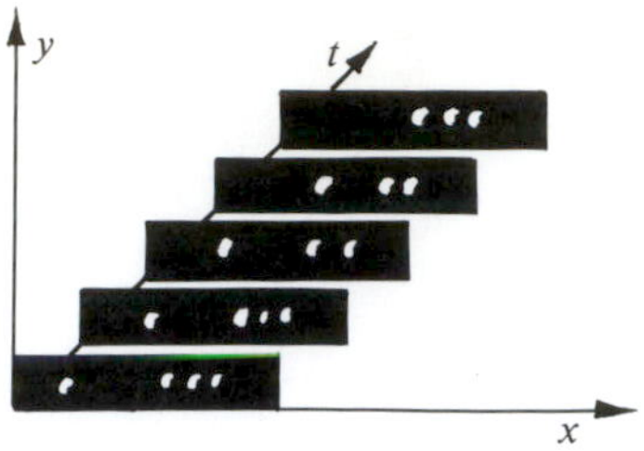

Fig. 4.27. Evolution of clusters of stable spots

4.4.3 Critical Phenomena

We shall briefly consider the behaviour of structures in inhomogeneous excitable media. Nonequilibrium media have typical inhomogeneities of two types:

a) smooth ones, that are determined, for example, by the inhomogeneity of the external field controlling the processes in the medium, and
b) random inhomogeneities caused by defects (unexcitable or distorted portions) that are on the average uniformly spread over the medium.

Let us consider examples in which the effects of these two inhomogeneities can be seen independently.

In excitable media with uniformly distributed defects the propagation speeds of excitation fronts will naturally slow down as the density of the defects increases. This slowing down is indeed observed. It turns out to have a critical behaviour and to be similar to the percolation phenomena. Results of experiments performed with our TV-analogue are presented in Fig. 4.28 [4.59].

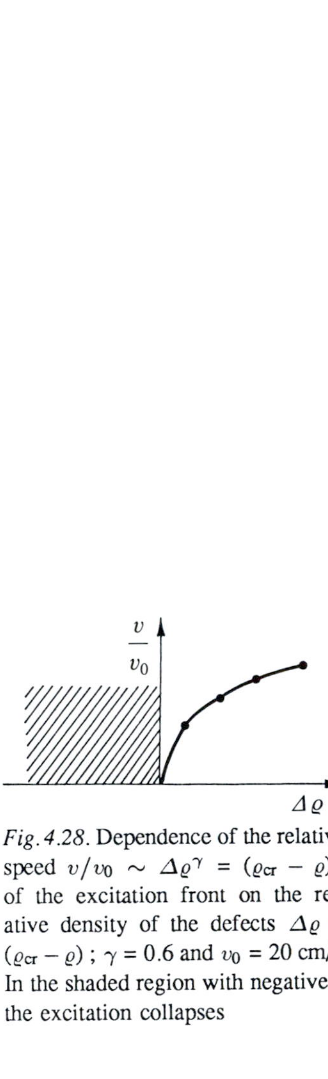

Fig. 4.28. Dependence of the relative speed $v/v_0 \sim \Delta\varrho^\gamma = (\varrho_{cr} - \varrho)^\gamma$ of the excitation front on the relative density of the defects $\Delta\varrho = (\varrho_{cr} - \varrho)$; $\gamma = 0.6$ and $v_0 = 20$ cm/s. In the shaded region with negative ϱ the excitation collapses

In particular, we found that there exists the critical value of defect density ϱ_{cr}, below which stationary propagation of excitation fronts (or switching from one state to another) occurs in an excitable medium with defects, as well as in a homogeneous medium. If the defect density is higher than the critical value, the excitation does not propagate, but instead collapses. Near the critical point ($\varrho \leq \varrho_{cr}$), the front speed changes according to the power law $\Delta\varrho^\gamma = (\varrho_{cr} - \varrho)^\gamma$, where the

[8] All the effects produced within the "monitor + camera + feedback" system are independent of the phase relations of the optical field (only the brightness of the screen enters (4.25) or (4.26)). However, similar phenomena – kinematic spiral waves, various dislocations and spatio-temporal disorder – may also be shown with coherent light where they are due to phase effects. For example, in a laser-pumped resonator with two-dimensional feedback and a liquid-crystal phase mask controlled optically [4.64]. In this case, the dynamic processes are also described by a nonlinear parabolic equation, however, not for the intensity but for the phase of the optical beam, which depends on the transverse coordinates and on time. For phase turbulence see as well [4.29].

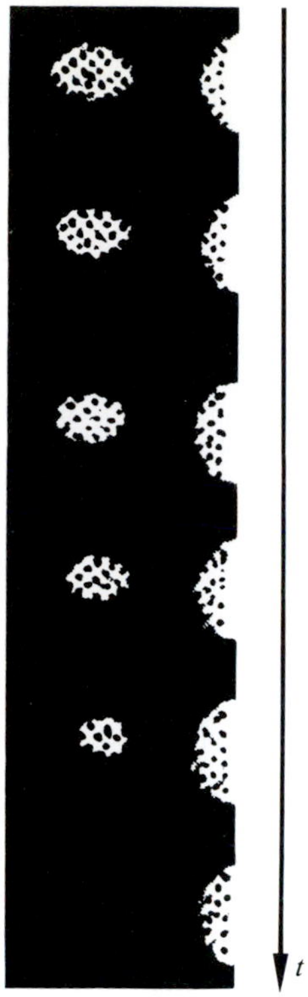

Fig. 4.29. Time evolution of a spatially restricted excitation in a medium with $\varrho > \varrho_{cr}$

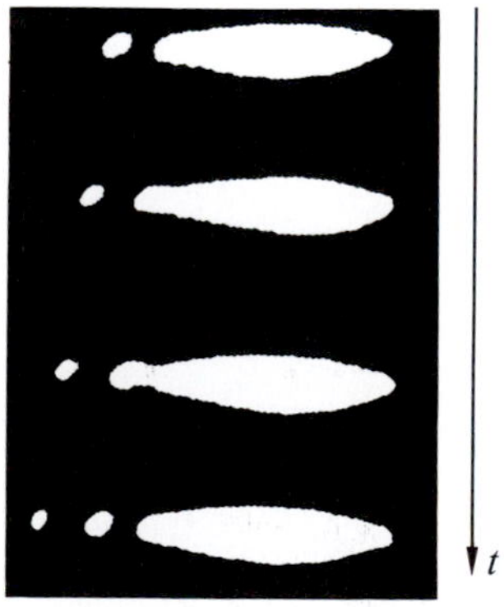

Fig. 4.30. Drops that separate periodically from a localized disturbance in a smoothly inhomogeneous medium with an inhomogeneity gradient exceeding the critical one

critical index $\gamma \in (0, 1)$ is determined by the degree of nonequilibrium of the system (or by contrast and brightness in experiments). Stationary fixed structures are observed when $\varrho = \varrho_{cr}$. A similarity is observed near the critical point – the increase in "contrast" is equivalent to the decrease in defect density. For $\varrho > \varrho_{cr}$ limited excitations collapse, see Fig. 4.29.

One of the most fascinating effects observed in a smoothly inhomogeneous excitable medium is the periodic rupture of "drops" off the initially stable excitation region.

At a certain critical value of the inhomogeneity gradient the front of the region of stationary excitation becomes unstable. The "drops" are formed at the nonlinear stage and then tear off, Fig. 4.30. The rate with which these drops are generated, grows with increasing inhomogeneity gradient.

4.4.4 Structures in Neuron-Like Media

Coming back to the problem of remembering and identifying images, it appears natural to use the TV-analogue for the modelling of associative memory. As a minimum, we will need a two-component nonequilibrium medium, having a "slow" and a "fast" component. Using a TV, we can realize such a medium in the form of screens of two monitors, each being connected to a closed circuit – "monitor + camera + amplifier + feedback". The principle of parallel identification of images is realized in such a nonequilibrium medium which is evidently also observed in Nature.

Indeed, recent neurobiological investigations show that brain functions such as recognition and identification of images result from the collective action of ensembles of neurons. In such processes the neurons function in parallel [4.65]. According to experiments, the excitation structures give birth to complicated spatial patterns in neuron networks during learning and identification. This excitation mosaic is detected by direct microelectrode and thermovision measurements. Unfortunately, we have so far failed to establish experimentally the nature and range of the neurons' interaction which is, undoubtedly, a most essential aspect for the construction of models of memory. We would like to recall that the number of neurons interacting in an excitation mosaic is $10^2 - 10^4$ and that it is continuously changing during the data processing by the brain. Therefore, we have to construct *a priori* models to solve then the inverse problem.

It appears most natural to assume the relation between the neurons to be isotropic and monotonically decaying as the distance between the interacting elements increases. However, the models of excitable media based on this hypothesis explain only the simplest collective effects observed in real neuron networks. They are, in particular, nonlinear waves of epilepsy-like activity and depression waves. However, these models do not describe the existence of localized centres of activity in the cortex.

We shall make use of our nonlinear experience gained in the analysis of localized convective structures (see for instance Sect. 4.3), and assume that the relation between the neurons in the network is not only nonlocal but also nonmonotonic depending on the distance between them. We will assume also that the neighbouring neurons excite each other, while those located rather far apart, inhibit other neurons.[9] The existence of localized centres of activity and many other phenomena connected with self-learning in neuron-like media can be explained within models taking into account this peculiarity of the relation between neurons. We will consider two models of this type.

Assuming that the control variable v (it must, evidently, be slow) changes only the excitation threshold of the variable u that reproduces information, we can give the equation for a model of a neuron-like medium:

$$\partial_t u = -u + \mathcal{F}\left[-\alpha + \int_S \Phi(\boldsymbol{\xi} - \boldsymbol{r})u(\boldsymbol{\xi}, t)d\boldsymbol{\xi} - \beta v\right] , \tag{4.28}$$

$$\partial_t v = \frac{\tau_u}{\tau_v}(u - v) . \tag{4.29}$$

In the simplest situations the neuron can only be in two states. Thus the function $\mathcal{F}(z)$ that characterizes the response of the elements in the medium to the stimulating action z, should be considered to be a step function. In this model the coupling function $\Phi(\boldsymbol{\xi})$ is fixed. Assuming that this function can be controlled, we arrive at more general models

$$\begin{aligned} \partial_t u &= -u + \mathcal{F}\left[-\alpha + \int_S \Phi(\boldsymbol{\xi} - \boldsymbol{r}, v_1, v_2)u(\boldsymbol{\xi}, t)d\boldsymbol{\xi}\right] , \\ \partial_t v_{1,2} &= \Psi_{1,2}(v_1, v_2, u) . \end{aligned} \tag{4.30}$$

Here the control variables v_1 and v_2 can determine, for example, anisotropy and range of the neuron coupling, respectively.

The medium described by (4.30) can, in particular, memorize the contours of pictures, distinguish the picture elements oriented in an assigned direction and has still further "intellectual" functions, see Fig. 4.31.

[9] Biologists refer to such interactions as lateral inhibition.

Fig. 4.31. Examples for the realization of "intellectual" functions of a neuron-like medium

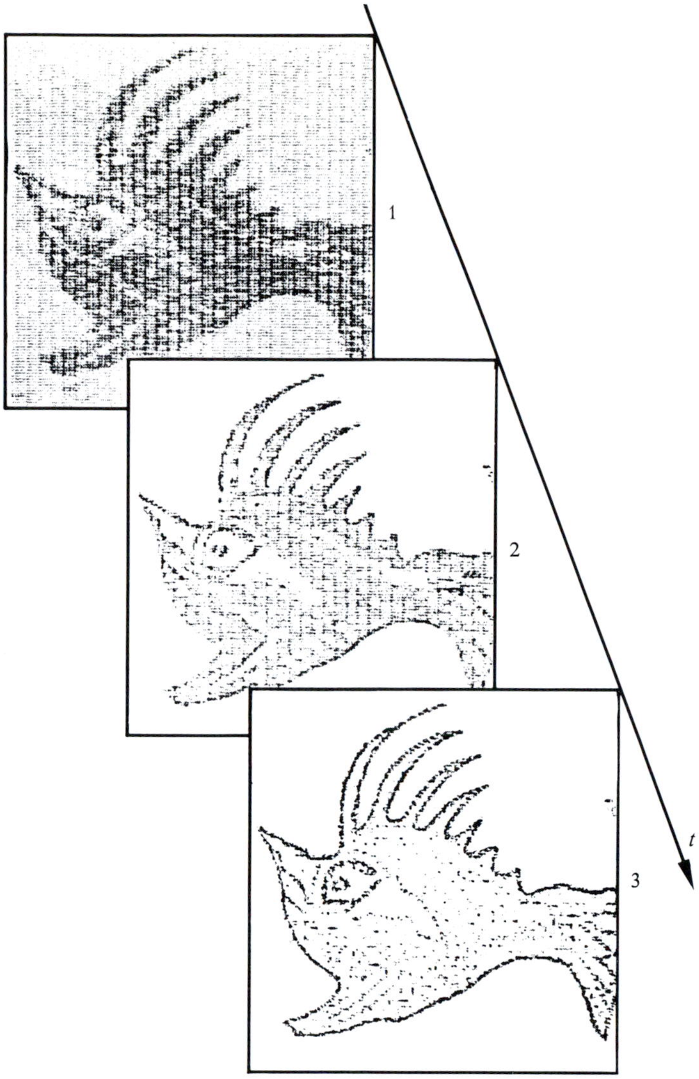

5. Turbulence

For almost a century classical physics has been attempting to solve the problem of turbulence. Nevertheless "we still do not understand how water flows" as Feynman put it. We believe that in the last ten years nonlinear physics has come a lot closer to understanding this phenomenon than expected twenty to thirty years ago.

5.1 Prehistory

Having investigated the deterministic chaos in finite-dimensional systems and the regular dynamics of nonlinear fields and media, we have approached the most challenging problem of modern classical physics – the problem of turbulence. Turbulence is an irregular, disordered behaviour of a continuous medium or field in time and space. The elucidation of the origin of this disorder is the fundamental problem of turbulence. Using the experience gained in our previous studies, we will now attempt to formulate alternative approaches to its solution. They are based on two paradigms:

1) Turbulence is the result of the amplification and nonlinear transformation of external or internal noises of the medium, flow, field, etc.
2) Turbulence is spatio-temporal deterministic chaos, that is the self-oscillations of a distributed system, that are stochastic in time and space.

The progress in theory and experiment made in the recent ten years seems to indicate that Nature prefers the second one. Hence, this chapter will be concerned with the theory of dynamic turbulence.

When speaking of the historical aspect of turbulence, it should be noted that presumably no other similarly widely observed phenomenon has attracted so much interest giving at the same time rise to violent discussions and contradictory arguments among physicists, mathematicians and engineers. This is explained not only by the extraordinary complexity of the system, but also by the radically different approaches to its understanding used by the researchers. Indeed, from the point of view of applications, turbulence is essentially the formulation of some effective equations (that are much simpler than the Navier-Stokes equations), which would enable us to calculate envelope resistance, effective heat and mass transfer and other properties of specific turbulent flows. Such equations may be written down on the basis of intuitive concepts and various semi-empiric hypotheses. The only important point is that they should predict answers to possible experiments. Remarkable results have already been obtained, see, for example [5.1]. However, the fundamental questions formulated above: "How can disorder arise in initially laminar (ordered) flow?" and "In what way does this disorder affect the average characteristics of the flow?" – are not answered by such an approach to turbulence. The averaged description does not enable us to understand whether the fluid elements intermingle and mix with each other as a result of uncontrolled pulsations of the flow running across a body, or whether they are due to instabilities inherent in the flow (of course, noise and fluctuations are also important, but in this case they simply act as a trigger initiating instability).

When you seek it,
you cannot find it.

Zen Riddle

These questions are essential for any turbulence, be it hydrodynamic, plasma, spin, or another type of turbulence. In other words, we are faced with the fundamental problem of finding out in what way a nonlinear field of arbitrary origin becomes disordered and how to describe this irregular motion.

In many aspects the evolution in the understanding of randomness is very similar in the theory of turbulence and in statistical physics. Indeed, the natural concept of small perturbations increasing as a result of developing linear instability as formulated by Osborne Reynolds could not explain the transition to turbulence in seemingly simple flows such as the Poiseuille or Couette flow between planes. According to the analysis of small oscillations within the Navier-Stokes equations, there exists no critical Reynolds' number at which the flow in a tunnel could become unstable. In contrast to that all experimentors know only too well that this flow becomes turbulent for $R \gtrsim 5000$. Such "failures" of the linear theory gave rise to many doubts as to the validity of deterministic Navier-Stokes equations for the description of turbulent flows. Karman and Taylor, for example, believed that, similar to the motion of gases, turbulence could be understood and interpreted only on the basis of a statistical approach. Their point of view was supported by the fact, which appeared quite apparent at the time, that the number of degrees of freedom participating in the motion is so large for a sufficiently large Reynolds' number, that only an average description of the flow is possible.

The discovery of deterministic chaos changed those concepts completely. However, the first attempts to explain a disordered chaotic motion of "water flow" in terms of pure dynamics had been undertaken long before, back in the "pre-chaotic era" by *Landau* (1944) [5.2] and *Hopf* (1948) [5.3].

They proposed models which realized the idea of successive complication of the flow due to a developing hierarchy of instabilities with incommensurate time scales. The nonlinear stabilization of these instabilities results in a flow, whose velocity field is the more disordered the larger the number of perturbations with incommensurate scales participating in its formation. If we construct the self-correlation function of the velocity field for such a flow, we will see that the correlations decay rapidly. Such a process exhibits a definite regularity only at times greater than the Poincaré recurrence time $T \sim e^{\alpha N}$ with $\alpha \sim 1$, where N is the number of perturbations with incommensurate frequencies ω_j ($j = 1, 2, \ldots, N$). Hopf constructed a simple dynamic system, with a solution possessing an infinite number of such incommensurate frequencies. In the phase space of the proposed flow successive bifurcations occur as the Reynolds number increases. First, a trivial equilibrium state gives birth to a limit cycle (the image of a single-frequency periodic flow), then the cycle becomes unstable and a bifurcation gives rise to an attractor in the form of an open winding on a two-dimensional torus. After the k-th bifurcation, an attractor in the form of an ergodic winding on a k-dimensional torus appears, etc.

Landau-Hopf's hypothesis played a significant role in shaping the concepts of the dynamical origin of turbulence – although their models (proposed more than forty years ago) proved to be inadequate. We now know that an attractor in the form of an open winding on a multi-dimensional torus (which is the image of turbulence within those models) is structurally unstable, i.e. it is destroyed for small variations in the parameters of the system. This implies that such a disordered flow (or, to be more exact, a complex flow with a great number of incommensurate frequencies in the Fourier spectrum) cannot be realized as a rule.

Nevertheless, the notion that the development of turbulence is determined by successive excitations (with regard to the Reynolds number) of still further degrees of freedom in the flow, proved to be quite correct.

The "new era" in the development of the dynamic approach to turbulence began with the discovery of strange attractors. One of the first strange attractors – the Lorenz attractor – was discovered in the investigation of a model of a hydrodynamic flow. The scenarios of the transition to chaos in finite-dimensional systems developed in the seventies, were also associated with the birth of turbulence. Subsequent experiments confirmed the onset of turbulence in closed cavities (convection in a cavity, the Couette-Taylor flow, and other closed flows) to be described by bifurcation sequences typical for the transition to chaos in simple dynamic systems, for example, one-dimensional maps. Moreover, on the basis of the ideas suggested by Takens, a strange attractor beyond the transition point was reconstructed directly from the observed data and the dependence of the dimension on the Reynolds' number was determined [5.4, 5], see Fig. 5.1. However, in spite of the remarkable success, there were also sceptic opinions such as "although admittedly discovered in water flow, finite-dimensional temporal chaos is not yet turbulence". We have to agree with this statement. Indeed, the spatial complexity is no less typical of turbulence than the temporal one.

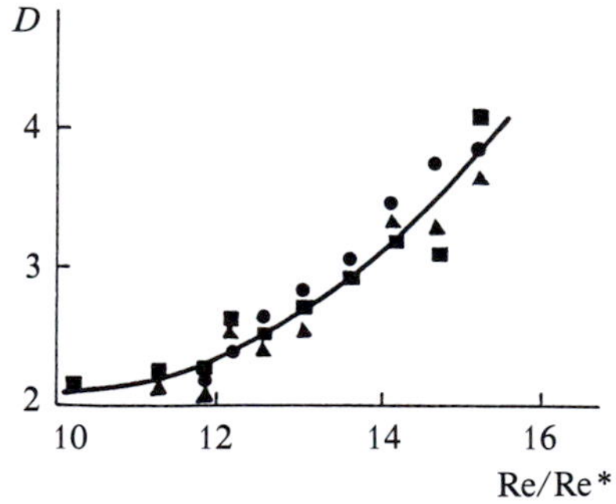

Fig. 5.1. Dependence of the dimension D of a strange attractor (modelling a turbulent Taylor-Cuette flow between rotating cylinders) on the Reynold's number $Re/Re^* = Re/Re_{cr}$

5.2 Basic Models of Dynamic Theory

In order to explain and describe onset and development of spatio-temporal disorder as observed experimentally, we need a theory of dynamic turbulence, which must at least allow for a description of the scenarios according to which spatio-temporal disorder develops with increasing supercriticality, for example the Reynolds' number, and corresponding estimates of the dimension; the elucidation of the relation of the dimension of chaos to the topological structure of the field; the determination of the parameters of turbulence that characterize its spatial inhomogeneity and allow for comparison with the magnitudes measured experimentally. It should be emphasized that the description of spatially inhomogeneous turbulence is of particular importance for the construction of a theory of the self-generation and development of disorder in flow systems such as hydrodynamic shear flows, see for example Fig. 5.2.

It often occurs in physics that the success of a theory depends on a lucky choice of the basic model (or models). For example, the natural hypothesis of the universality of the ways in which spatio-temporal chaos develops or of the properties of individual structures related to it, enables us to consider elementary models that allow for sufficiently thorough investigations and are at the same time rather typical and complete (in terms of the diversity of the phenomena described). Be-

a

b

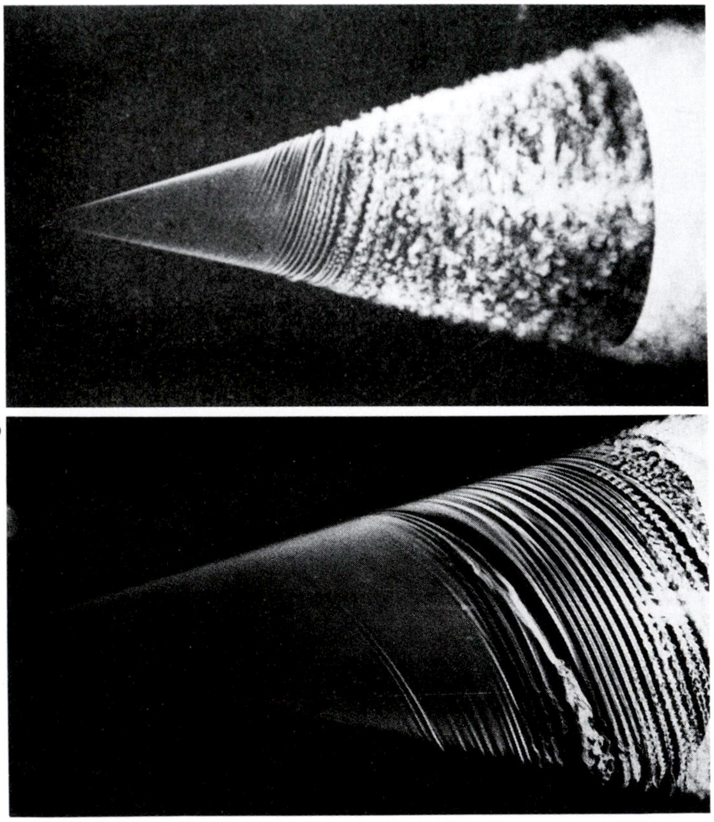

Fig. 5.2. Pattern of a vortex structure due to the boundary layer flow on a rotating cone

sides, these models should arise from approximations to the equations of motion for initial fields, for example, the Navier-Stokes equations.

We have encountered many models of this type in the theory of structures in nonequilibrium media, in particular, the generalized Ginzburg-Landau equation (GGLE)[1]

$$\partial_t u = u\left[1-(1+i\beta)|u|^2\right] + \kappa(1-ic)\Delta u \ . \tag{5.1}$$

Applied to hydrodynamics this equation describes, for example, the amplitude $u(x, y, t)$ of the complex $\exp[i\boldsymbol{kr}]$ velocity field for Rayleigh-Bénard convection, see *Newell et al.* [5.6]. For small supercriticality, the coefficients in (5.1) can be considered to be real and (5.1) will be written in the gradient form (4.10). Whence it follows that the field modelled by (5.1) with real coefficients can only be in a static state as $t \to \infty$. Depending on the parameters the number of such states having different distributions of $u(x, y)$, may be fairly large.

In the general case of complex coefficients, (5.1) describes both the time-periodic restructuring of a two-dimensional field structure and spatio-temporal chaos, i.e. dynamic turbulence.

Equation (5.1) holds when the ground state loses its stability softly. In the case of hard instability, (5.1) has to be exchanged against

$$\partial_t u = -u + (\beta + i\beta')|u|^2 u - (1-i\gamma)|u|^4 u + (1-ic)\Delta u \ . \tag{5.2}$$

With a convective term of the form $V_0\partial_x u$ these models describe many phenomena in flow systems (e.g. in shear and boundary layers, etc.). The respective extension of (5.1) is

$$\partial_t u + V_0\partial_x u = u\left[1-(1+i\beta)|u|^2\right] + \kappa(1-ic)\Delta u \ . \tag{5.3}$$

Nonequilibrium media have a tendency to 'structuring'.

In the previous chapter we have seen that nonequilibrium media have a tendency to "structuring". Localized patterns (self-structures) emerge in many instances. A relative autonomy of their dynamics allows us to consider the nonequilibrium medium (or field) as an ensemble of coupled structures, and we can go over from partial differential equations to difference equations for the parameters of coupled structures. Then restriction of our description of the interaction of excitations in neighbouring structures to linear coupling yields [5.10]

$$\partial_t u_j = \Phi[u_j, \varepsilon] - \gamma[u_j - u_{j-1}] + \kappa[u_{j+1} - u_j] + \xi\partial_{yy}u_j \ . \tag{5.4}$$

Here $\Phi[u_j, \varepsilon]$ describes the dynamics of an individual structure and the derivative $\xi\partial_{yy}u_j$ accounts for the fact that a localized structure (element of the chain (5.4)) may be extended along one of the coor-

[1] In 1950, V.L. Ginzburg and L.D. Landau constructed a phenomenological theory for the destruction of superconductivity by the magnetic field. Within this theory they derived the equations for the order parameter and vector potential [5.7]. At the end of the sixties those equations were generalized to the nonstationary case [5.6, 8, 9] and may be considered as fundamental equations of the theory of nonequlibrium media and nonlinear physics in general.

dinates, for example vortices in boundary layers and Taylor vortices, spirals on rotating bodies (see Fig. 5.2) and so on.

A similar model can be constructed for a two-dimensional ensemble of structures [5.11, 12]:

For further illustrations on the solutions of (5.5) see Chapter 6: Fig. 3 and Plates 9, 14, 29 and 63

$$\frac{du_{jl}}{dt} = u_{jl} - (1 + i\beta)|u_{jl}|^2 u_{jl} + \kappa(1 - ic)(u_{j,l+1} + u_{j,l-1} + u_{j+1,l} + u_{j-1,l} - 4u_{j,l}) \quad (5.5)$$

with $l, j = 1, 2, \ldots, N$ and $N \gg 1$. Within such models, it is convenient to assume N to be arbitrarily large but finite. Naturally, (5.5) has to be supplemented by boundary conditions.

It is important that "chain" or "lattice" models yield at least a qualitative description of the turbulence of structured fields and strongly nonlinear structures. For example, a chain of sine-Gordon solitons in the field of a harmonic wave is in the presence of dissipation described by

$$\partial_t u_j = \varepsilon u_j - \beta |u_j|^2 u_j + \kappa(u_{j-1} + u_{j+1} - 2u_j) , \quad (5.6)$$

where u_j stands for the displacement of the j-th soliton with respect to its equilibrium state in the wave potential [5.11].

5.3 Turbulence and Structures in Two-Dimensional Fields

5.3.1 Experiments

We will first present results of computer and laboratory experiments, so as to convince our reader that the "structural" approach is quite natural in the investigation of turbulence. We shall restrict ourselves to two-dimensional fields.

The behaviour of the field, in particular for $t \to \infty$, may be studied as a function of the degree of nonequilibrium within model (5.1) or its discrete (lattice) analogue (5.5). The parameter $r = 1/\kappa$ plays the role of the Reynolds number. For the spatial field distributions corresponding to the regimes of interest see Table 5.1 and Plates 14 and 29 of Chap. 6 *Nonlinear Physics*.

If the imaginary part of the coefficients is small, as compared to the real one, then the field is an ensemble of weakly interacting slowly moving long-lived structures. Formally, this is explained by the fact that our model is nearly a gradient one for almost real coefficients.

The structures described by (5.5) may have different topologies, and they transform into one another as supercriticality increases. In particular (see Sect. 4.3), the increase of supercriticality due to the loss of stability leads in the simple periodic lattice of coupled orthogonal standing waves to a lattice of spiral pairs, see Plate 29 (right) of Chap. 6 *Nonlinear Physics*. As supercriticality grows (κ diminishes), the regular spatial field pattern in the form of a periodic lattice of

Table 5.1. Classification of types of motion according to the parameter κ

κ		Type of motion for $\beta = c = \sqrt{3}$
		Stable homogeneous equilibrium
102	—	
		Stationary large-scale structures: rolls, concentric vortices, spiral pairs
80	—	
		Quasi-stationary structures: "breathing" spiral pairs, modulated vortices, etc.
35	—	
		Chaotic dynamics of structures – structural turbulence
5	—	
		Developed high-dimensional spatio-temporal chaos: "boiling turbulence"
0.25	—	
		Developed modulation chaos against the background of lattice oscillations. Hierarchy of excitations
1/16	—	
		Stable periodic dynamics with anti-phase oscillations of neighbouring lattice elements

spirals is destroyed, the spirals become more and more independent – they move in the field of each other, first in an ordered manner and then chaotically. Dynamical turbulence is being formed, see Plate 29.

A further increase of supercriticality leads even to the destruction of the spirals and the spatio-temporal chaos gets more complicated. Turbulence develops further and further. We will describe the degree of turbulence, as $r = 1/\kappa$ increases, through the κ-dependence of the dimension of the limiting stochastic set.

However, we shall first describe a laboratory experiment which makes it possible to observe dynamic turbulence in a homogeneous isotropic medium. We will be concerned with the parametric excitation of waves by a homogeneous oscillating field [5.13]. The generation of pairs of counter-propagating waves at a frequency close to one half of the pump frequency (see Fig. 5.3) is of interest to many branches of physics: in connection with the excitation of Langmuir waves in plasmas, spin waves in ferromagnetics, waves on the surface of a capillary fluid, liquid dielectric in an electric field, or a ferrofluid in an alternating magnetic field. A rather unexpected manifestation of parametric instability is the emergence of wave patterns on the surface of a molten metal heated by modulated ion beams, Fig. 5.4 [5.14].

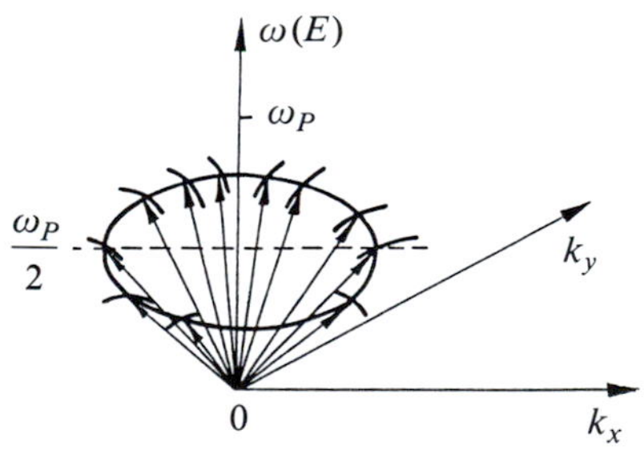

Fig. 5.3. Dispersion law illustrating the parametric excitation of a pair of capillary waves

The parametric excitation of surface waves on water in a variable gravity field is a most pictorial phenomenon of this type. It was investigated by Michael Faraday back in 1831. One and a half centuries later, the interest in such experiments was revived in connection with the investigation of spatio-temporal chaos of structures. In this particular case, the structures are the cells of capillary lattices, various localized defects and singularities against the background of this lattice, etc.

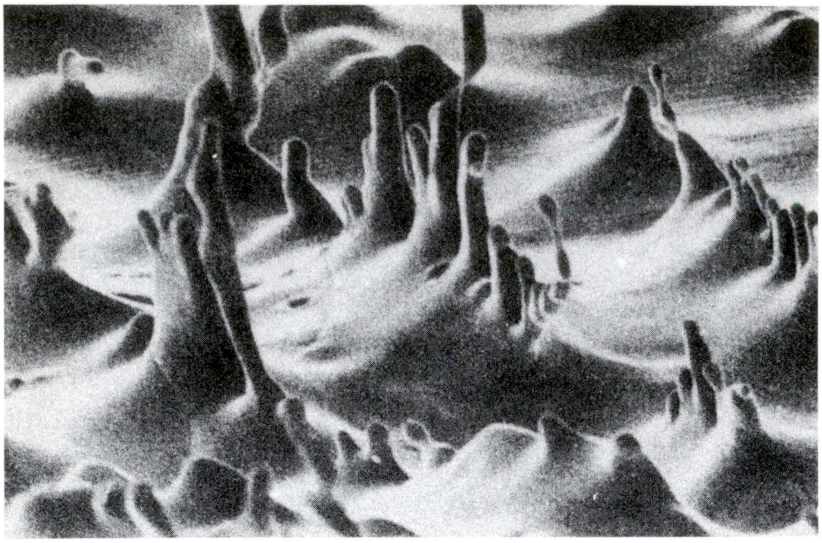

Fig. 5.4. "Parametric ripples" on the surface of a molten metal

Thus, we will be interested in the birth of turbulence on the surface of a parametrically excited fluid layer, Fig. 5.5. A horizontally placed membrane oscillates homogeneously in space at a given frequency $f = f_0$. Pairs of parametrically coupled capillary-gravity waves appear in the oscillating gravity field on the surface of the fluid layer on the membrane. For not too large vibrational (or pumping) amplitudes these wave pairs form an exceptionally regular spatial lattice with square cells. The same picture was observed by Faraday. However, if we increase the oscillation amplitude of the membrane (i.e. the supercriticality parameter), a quite unexpected phenomenon is observed – spatio-temporal disorder (turbulence) emerges but the individual cells do not break, see Fig. 5.6. The transition from a discrete to a continuous spatial spectrum of surface waves corresponds to the transition from a regular to a disordered arrangement of the cells.

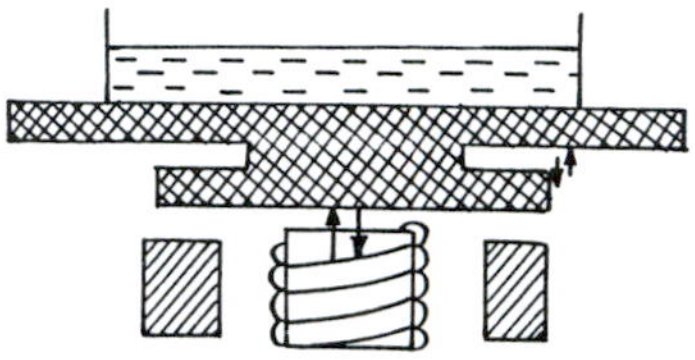

Fig. 5.5. Experimental set-up for observing parametric excitations of capillary waves:
silicon oil in a bowl is placed on a membrane oscillating up and down with the frequency f_0 = 112 Hz; the coil represents an electromagnet driving these oscillations

There are various scenarios of the birth of a continuous spatial spectrum. They depend on the way in which supercriticality is increased. Three types have been observed experimentally and investigated in more detail. If the oscillation amplitude of the membrane is increased adiabatically, one-dimensional stochastic modulation of square lattices is seen; Fig. 5.7a. In this case, for a complex modulation amplitude, it is possible to derive a parametric version of the Ginzburg-Landau equation. The dynamic theory constructed on the basis of this equation [5.15] describes (even quantitatively) the transition to turbulence in this spatio-temporal problem. A sufficiently abrupt increase of the pumping amplitude yields a different transition to turbulence which will be due to the birth of dislocations, Fig. 5.7b (see also Fig. 4.13a,b). The mechanism of their formation is related to the competition of modulation waves having close spatial periods. Such a competition (at the boundary of the transition of one modulation type to another) results in topological singularities of the nonlinear field. The chaotic dynamics of these interacting singularities is the turbulence born within this scenario.

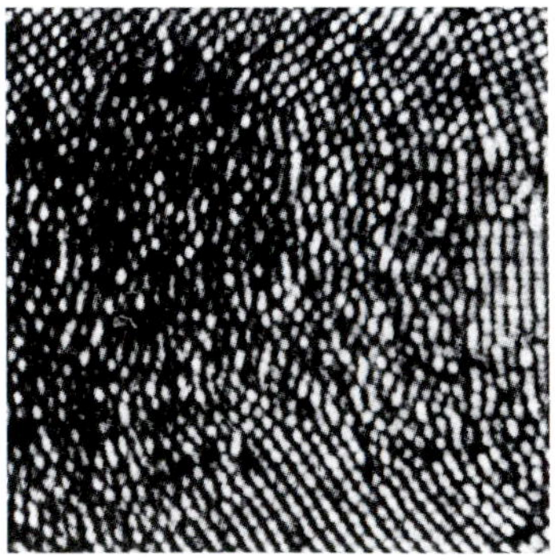

Fig. 5.6. Chaos of capillary cells

a
b
c
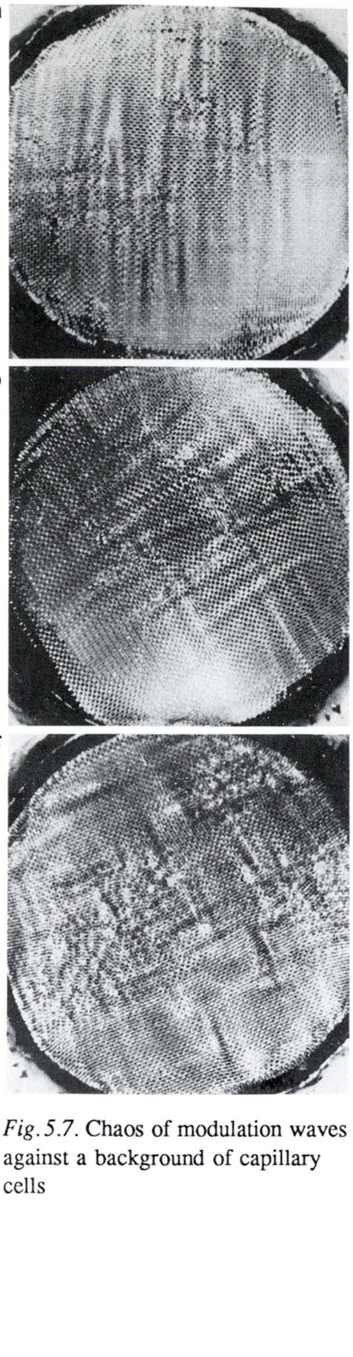

Fig. 5.7. Chaos of modulation waves against a background of capillary cells

Finally, in an intermediate case changes of the amplitude (or specific initial conditions) give rise to a quadratic modulation lattice with a period incommensurable with that of the quadratic cells. After the destruction of this quasi-periodic structure one observes also turbulence.

Given a sufficiently large supercriticality or vibration amplitude (corresponding to an "overload" of about 5 to 6 g), the properties of the observed turbulence will no longer depend on the way in which it has been established – turbulence is now spatio-temporal dynamic chaos in an ensemble of interacting capillary cells, Fig. 5.6.

5.3.2 Development of Turbulence and Multi-Dimensional Attractors

To develop a theory for such a chaos we consider the lattice model (5.5) with the periodic boundary conditions[2]

$$u_{jl}(t) \equiv u_{j+N,l+N}(t) \; . \tag{5.7}$$

As mentioned, modulation waves appear against the background of the homogeneous state near the boundary where it loses its stability. The developing secondary instability results in the complication of modulation – ensembles of spirals are born, which are destroyed with the further increase of $r = 1/\kappa$ leading to the onset of turbulence. To estimate the extent, to which turbulence has developed within systems of the form (5.5), (5.7), it is natural to determine the dimension and entropy of the corresponding stochastic sets (see Sect. 3.4.2),

$$D_\lambda = m + d \quad \text{and} \quad H_\lambda = \sum_{j=1}^{k} \frac{\lambda_j}{\lambda_1} \quad \text{with} \quad (\lambda_k \geq 0 \quad \text{and} \quad \lambda_{k+1} \leq 0).$$

The entropy H_λ depends only on the number of unstable directions on the attractor and on the relative velocity with which neighbouring trajectories spread along these routes. The magnitude of H_λ is always smaller than the stochastic set dimension D_λ which characterizes the number of normal variables needed for the description of the established dynamic turbulence.

Assuming that the established turbulence is on the average homogeneous, i.e.

$$\left\langle |u_{jl}^0|^2 \right\rangle_t = |u^0|^2 \quad \text{and} \quad \left\langle (u_{jl}^0)^2 \right\rangle_t = (u^0)^2 \; ,$$

(u_{jl}^0 denotes the pulsation background of the strange attractors) and the average intensity pulsations are small

$$\left\langle |u_{jl}^0|^2 - |u^0|^2 \right\rangle_t = z_{jl} \quad \text{and} \quad \left(z_{jl} \right)^{1/2} \ll |u^0|^2 \; ,$$

[2] Computer experiments modelling system (5.5) considered ensembles with the total number of elements $N \times N = 128 \times 128$.

we are in the position to calculate explicitly the characteristic exponents of the motion on the strange attractor and to find their dependence on the parameter κ [5.12, 16]. In particular, we obtain at $\kappa < 1/4$ for D_λ the relation

$$D_\lambda = \frac{N^2[1 + (1 - 4\kappa)]}{\left[\left(1 + \beta^2\right)^{1/2} + 2\right] |u^0|^2 - 1}. \tag{5.8}$$

... systems with a similar strange behaviour are frequently encountered. These systems display a so-called chaotic time evolution. Indeed, the irregular behaviour of a larger and larger class of phenomena can be described with a relatively small class of mathematical objects, each of them specifying an iteration procedure that, despite its deterministic nature, produces time evolutions which are really unpredictable.

D. Ruelle
"Chaotic Evolution and Strange Attractors"
(Cambridge University Press, Cambridge 1990) p. ix

We can see that the dimension of dynamic turbulence grows, as the supercriticality increases (i.e. as κ decreases). Direct computations show that in this process new independent excitations are involved in the turbulent motion. In our case, they are stationary waves of the form

$$u_{jl}(t) = A_n \exp[i(\omega_n t + j\Theta_n + l\Psi_n)],$$

where ω_n, Θ_n and Ψ_n are related by the dispersion equation that follows from (5.5). Naturally, all of them are unstable.

Unlike the transition to chaos, the scenarios of its development with increasing supercriticality are much more difficult to identify in a computer experiment. Nevertheless, it was possible to study within model (5.5) the two most typical mechanisms for the development of dynamic turbulence, whose image is a multi-dimensional strange attractor. The "accumulation" of instabilities for varying parameters is a precondition facilitating the existence of such an attractor in the phase space of the system, describing the spatio-temporal dynamics of the nonlinear field. In other words, a multi-dimensional strange attractor cannot emerge all of a sudden in a region of the phase space, without bifurcations or trajectories with a separatrix manifold. It follows that there must be two ways giving rise to multi-dimensional strange attractors, corresponding to the hard and soft onset of dynamic turbulence, respectively.

The first way is as follows: For $\kappa > \kappa^*$ a nonattracting stochastic set and a low-dimensional attractor (e. g., equilibrium or stable cycle) exist in the phase space; for $\kappa < \kappa^*$ the low-dimensional attractor disappears and the "former" nonattracting set is incorporated into the multi-dimensional strange attractor with hard onset. For small κ this scenario is realized in the discrete model (5.5) [5.17].

Another possibility is connected with a gradual increase (as $r = 1/\kappa$ grows) of the number of unstable directions for the attractor trajectories. In fact, (5.8) corresponds to this situation. As κ diminishes, the Lyapunov exponent for typical attractor trajectories slowly passes through zero which results in the growth of the attractor dimension.[3]

In the spatio-temporal picture, the increase of the dimension of dynamic turbulence is accompanied by the complication of the spatial pattern of the field, in particular, by an increase of the number of

[3] Numerical experiments have demonstrated the basic mechanisms of the increase of attractor dimension – period doubling of the saddle cycle in the attractor and birth and subsequent destruction of the saddle torus from the saddle cycle.

interacting structures – defects, spirals, etc. – in the lattice. Obviously, the number of these structures is restricted by the number of lattice elements. For large supercriticality (for instance, when $r = 1/\kappa \geq 0$ in (5.5)), spatio-temporal chaos is established in the discrete lattice. In this process short-wave excitations play the main role. Within limits, these are "π-π-oscillations", corresponding to out-of-phase oscillations of neighbouring lattice elements. A very fascinating spatio-temporal chaos is born in this case – it is a kaleidoscope of pictures never repeated in time. Its spatial structure is determined by the number of lattice elements (for an analogy see Plates 57–62 of Chap. 6 *Nonlinear Physics*).

We would like to emphasize that it is very difficult to order, or synchronize such a chaos. We shall explain this remark using as an example a lattice of coupled pendulums [5.18] with

$$\begin{aligned}\ddot{\varphi}_{jk} + \mu\dot{\varphi}_{jk} + \sin\varphi_{jk} &= F(\omega t, j, k)\\ &+ \zeta(\varphi_{j-1,k} + \varphi_{j+1,k} + \varphi_{j,k-1} + \varphi_{j,k+1} - 4\varphi_{jk}) \,, \end{aligned} \tag{5.9}$$

with $j, k = 1, 2, \ldots, N$; $\varphi_{jk} \equiv \varphi_{j+N;\, k+N}$ and the external force $F(\omega t, j, k) = A \sin\omega t \cdot f(j, k)$ which is periodic in time and directed along the spatial coordinates. Given a sufficiently great amplitude of the external field, all oscillators are out of phase and in no way repeat the spatial distribution of the external field, see Plate 9 of Chap. 6 *Nonlinear Physics*.[4]

However, it is now realized that if the quasi-modes interact strongly, even a small number (greater than two) of them can give rise to a pseudo-random behaviour. To all intents and purposes the behaviour is random but arises from a small set of coupled deterministic equations. Such a phenomenon is called chaotic.

E. Infeld, G. Rowlands "Nonlinear Waves, Solitons and Chaos" (Cambridge University Press, Cambridge 1990) p. 25

Obviously it is of interest to study the way in which bifurcations evolve into such a chaos in real Euclidian space and not in phase space. The development of the theory of bifurcations for such spatial patterns is still in its initial stage. In particular, bifurcations of the "commensurability–incommensurability" and "commensurability–chaos" types have been observed in lattices of pendulums excited by homogeneously oscillating field (for sufficiently great N and "dissipative" boundary conditions).

5.4 Spatial Evolution of Turbulence

5.4.1 Flow Dimension

If the theory of dynamic turbulence is really valid for the description of flowing water then it must describe the onset and spatial evolution of "disorder" downstream. Apparently, this is of particular significance for shear flows (boundary layers, submerged jets, wakes, etc.). In what way is the dynamic chaos self-generated along the flow? Are the scenarios of its appearance similar to the scenarios observed in simple systems with varying control parameter(s)? How does chaos behave

[4] Note that such problems have a significant interest in applications related to the investigation of the nonlinear dynamics of lattices of Josephson junctions in various antenna lattices, synchronization networks, power networks, etc.

infinitely far downstream? – These are the main problems to be answered by theory.

The mind must always be in the state of 'flowing', for when it stops anywhere that means the flow is interrupted and it is this interruption that is injurious to the well-being of the mind. In the case of the swordsman, it means death.

Takuan

As in the analysis of turbulence in isotropic media, it is in the investigation of spatio-temporal chaos in flow systems natural to employ the (fractal or correlation) dimension as the characteristic of the flow. However, in suitable systems the turbulent fields are strongly inhomogeneous so that the previously introduced parameters C or D have to be modified [5.10, 19].

We assume that the time series of the field $u(x,t)$ measured by a detector is at each point along the flow described by a different dynamic system $\hat{G}_x\{u(t)\}$. Its motion in n-dimensional phase space is described by the trajectory (here the index x does not stand for a partial derivative)

$$\begin{aligned}\boldsymbol{u}_x(t) &= \{u_x(t), u_x(t+\tau), u_x(t+2\tau), \ldots, u_x(t+(n-1)\tau)\}\\ &= \{u_1, u_2, \ldots, u_n\} \,.\end{aligned}$$

As before (see Chap. 3), we will calculate the dimension with the aid of the correlation integral; but now we will assume it to depend on the coordinate along the flow

$$C_x(\varepsilon) = \frac{1}{N_0^2} \sum_i \sum_j \Theta\left[\varepsilon - \|\boldsymbol{u}_i - \boldsymbol{u}_j\|\right] = \frac{N_\varepsilon(x)}{N_0} \,. \tag{5.10}$$

For ε lying in the finite interval $[\varepsilon', \varepsilon'']$, the correlation integral is approximated by

$$\nu_\varepsilon(x) = \frac{d \log[C_x(\varepsilon)]}{d[\log \varepsilon]} \,. \tag{5.11}$$

The function $\nu_\varepsilon(x)$ is the flow dimension we are interested in. The validity of this characteristic is confirmed by direct experiments, for instance, with a turbulent boundary layer on a plate. The experimental setup and the flow visualization are presented in Figs. 5.8, 9 and a plot of the dimension versus the coordinate is shown in Fig. 5.10 [5.10, 19]. We see that the dimension of the time series is constant. In particular, it does not depend on the coordinate in the region of the flow where stable two-dimensional vortex structures exist. The dimension grows downstream, at the point where three-dimensional vortices appear against a background of two-dimensional ones. A further increase

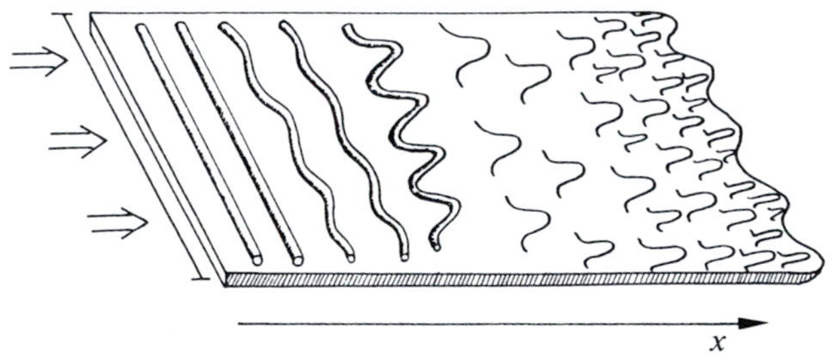

Fig. 5.8. Schematic picture of structures in boundary layer on a plane plate (flowing in the direction of the arrows). The ribbon on the left-hand-side excites the incident current and leads to two-dimensional vortices

Fig. 5.9. Visualization of the periodic excitations of a flow of a restricted fluid layer across a plate

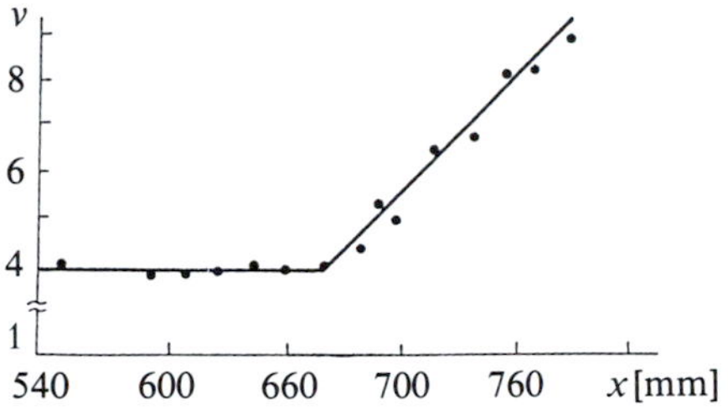

Fig. 5.10. Change of correlation-dimension ν across the boundary layer is illustrated

of the dimension downstream indicates that the motion involves more and more dynamic degrees of freedom of the flow. This process is, essentially, the development of turbulence along the boundary layer which enables us to construct its dynamic model.

5.4.2 Spatial Bifurcations

When constructing the model, we shall bear in mind that the developing primary instability in the flow system results in the formation of structures, whose collective dynamics will subsequently lead to the birth of turbulence, see Fig. 5.2. Then, in the simplest formulation we can write

$$\frac{du_j}{dt} = \Phi(u_j, \delta) + \gamma(u_j - u_{j-1}) + \kappa(u_{j+1} + u_{j-1} - 2u_j) , \qquad (5.12)$$

with $j = 1, 2, \ldots$, see (5.4). Here $\Phi(u_j, \delta)$ describes the dynamics of a discrete element of the medium, the quantity γ denotes the action of one element on another downstream and κ determines the feedback upstream. Equation (5.12) is supplemented by boundary conditions like $u_0(t) = Ae^{i\omega t} + c.c.$ which correspond to a time-periodic excitation of the flow.

Ignoring diffusion in (5.12), the birth of spatio-temporal chaos along the flow reduces to the problem equivalent to the onset of a strange attractor in a chain with interacting elements possessing regular individual dynamics. A typical feature of such a "discrete flow" is the existence of spatially inhomogeneous stationary solutions $u_j(t) = \bar{u}_j$ determined by $\bar{u}_{j-1} = (\Phi(\bar{u}_j, \delta) + \gamma\bar{u}_j)/\gamma$. For such a stable spatial distribution the motion continues to be laminar – no chaos is developed. However, if the solution becomes unstable for $j > j^*$, a more complicated motion is established. In general, it is characterized by the dimension. This motion, for example a quasi-periodic one, may become unstable for $j = j^*$. Thus the motion along the flow will be more complicated until chaos sets in at some j_{cr}. More rigorously, this implies (under the boundary conditions under consideration) the existence of an attracting limiting set – a strange attractor – in the phase space of the system of $j = j_{\text{cr}}$ dynamic elements. The parameter controlling the transition to dynamic chaos along the chain is within the considered theory $\bar{u}_j$; it determines the inhomogeneous spatial distribution of the field. In the problem of determining the characteristic Lyapunov indices along j, this parameter plays a role similar to the one of the control parameter δ. A natural consequence of this is the hypothesis of the similarity of the scenarios for the transition to chaos in a point system with the variation of δ and for the birth of dynamic turbulence in a flow system along j.

Changing the dynamic properties of the point element of the medium we can describe various scenarios of the spatial evolution of turbulence (through period doubling, quasi-periodicity, intermittency, and others). An intriguing spatio-temporal similarity has also been observed, namely scaling [5.20].

The development of chaos downstream corresponds in the theory to an increase of the dimension of the representation along the coordinate. Figure 5.11 [5.21] shows the plot for the dimension of the motion in the deterministic system considered – a chain of directionally coupled structures, each of which exhibits in an autonomous regime only periodic oscillations. From some point on the chain onwards, the dimension of the motion grows until saturation or "stabilization of the dimension" occurs. This phenomenon is typical for a chain of nondestructing structures and is related to the effect of stochastic synchronization of irregular oscillations in dissipative systems [5.22].

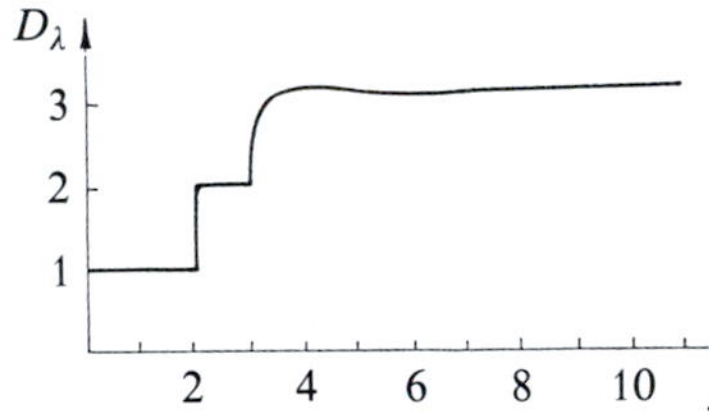

Fig. 5.11. Illustration of the "stabilization of the dimension" D_λ in a uni-directional chain of generators (j denotes the number of the respective generator)

5.5 Discussion

Phenomenological models of the form (5.1–5) prove to be very fruitful for an understanding of the mechanisms of turbulence and for the qualitative description of turbulent flows. However, for the quantitative characteristics of turbulent motion we have to refer to the basic three-dimensional field equations,[5] in particular, to the Navier-Stokes

[5] The exception are the characteristics of the transition to chaos near the critical point. The universal laws, that are determined by the type of critical behaviour only, are in this case in force, see Chap. 3

equations. In view of the exceptionally difficult analysis, there are only rather few results available, most of which are numerical ones.

Thus, computer experiment with the complete equations of hydrodynamics [5.23] in the form

$$\begin{aligned} \frac{1}{\sigma}\left[\partial_t \nabla^2 \Psi + \partial_{(x,z)}\left(\Psi, \nabla^2 \Psi\right)\right] &= R_T \Theta_x - R_S S_x + \nabla^4 \Psi \ , \\ \partial_t \Theta + \partial_{(x,z)}(\Psi, \Theta) &= \Psi_x + \nabla^2 \Theta \ , \\ \partial_t S + \partial_{(x,z)}(\Psi, S) &= \Psi_x + \tau \nabla^2 S \ , \end{aligned} \tag{5.13}$$

Two routes to chaos are through a period doubling sequence and via the destruction of quasi-periodic motion.

describing the diffusion of temperature and salinity in a horizontal fluid layer heated from below, demonstrate the transition to turbulence via a period doubling sequence. Here σ is the Prandl number, $0 < \tau < 1$ is the relation of impurity diffusion to thermal diffusion, R_T and R_S are the thermal and impurity Rayleigh numbers, respectively, $\Psi(x, z)$ is the current function, and Θ and S are, respectively, the deviations of temperature and impurity density from equilibrium. This system was solved by the finite difference method under the boundary conditions $\Psi = \Psi_{zz} = \Theta = S = 0$ for $z = 0, 1$ and $\Psi = \Psi_{xx} = \Theta_x = S_x = 0$ for $x = 0, \Lambda$. Another scenario of the transition to chaos – through the destruction of quasi-periodic motion – was observed in numerical experiments with the complete Navier-Stokes equations. Calculating the flow around a Zhukovsky profile ($R_e \sim 10^3 - 10^4$), an increase of the angle of attack was noted to lead to the transition from a stationary flow (at small angles) to a periodic one (at large angles), a quasi-periodic one and eventually to a turbulent flow. Thus, we can now state that the transition to turbulence following scenarios as observed in low-dimensional dynamic systems, is also realized within the basic equations of hydrodynamics.[6]

Returning to the role of structures in turbulence, we would like to emphasize that far beyond the transition point, i.e. at great Reynolds' numbers, turbulence is characterized not only by the complication of the collective dynamics in the ensemble of structures, but also by the presence of the structures with different parameters and properties which transform into each other during interaction. In this connection the computer analysis of the two-dimensional turbulence as described by the Navier-Stokes equations with periodic boundary conditions [5.25] should be recalled: it was found that the dimension of the stochastic set calculated by the aid of the Lyapunov exponents, is usually much smaller than the number of elementary (harmonic) excitations producing the flow. This fact is readily interpreted. The chaotic dynamics of the flow is determined by a small number of independent (in this case large-scale) excitations only, all the others are rigidly coupled to it so that they form coherent structures. In connection with the analysis of turbulence as modelled by the two-dimensional Ginzburg-

[6] Strictly speaking, problems that demand the grid method for their solution are as well related to finite-dimensional systems. However, because the number of elements is extremely large ($\gtrsim 10^6$) it is certainly tolerable to associate these results with the Navier-Stokes equations.

Landau equation we have already discussed similar results. Thus, the dimension of chaotic motion – turbulence – is seen to correlate with the number of the interacting structures (spiral waves, defects of the wave lattice, etc.).

Of utmost interest is the task of establishing a relation of the dimension of turbulence for great Reynolds' numbers to the number of modes in the inertial interval. So far only singular results are available for this problem, see [5.26]. However, it could well be that an explanation of the self-similarity of turbulence in terms of nonlinear dynamics may be obtained in the analysis of multi-dimensional stochastic sets that vary specifically when widening the inertial interval.

We would like to add that in real space the structure of turbulence is extremely complicated, moreover, recent experiments have shown it to be fractal, see Plates 40–42 of Chap. 6 *Nonlinear Physics*. Because of the significant inhomogeneity in the distribution of the scales of turbulence in space (abrupt boundaries between turbulent and nonturbulent fluids, formation of singularities – collapses, etc.), we have to consider sophisticated models that take into account the interaction of spatially separated dissipative and "conservative" structures.

6. Nonlinear Physics – Chaos and Order

"Myself when young did eagerly frequent
Doctor and Saint, and heard great argument
About it and about: but evermore
Came out by the same door as in I went."
Rubáiyát of Omar Khayyám 1859
English by E. Fitzgerald

6.1 The Where and the How

In science as in daily life we run into questions the answers to which are so sophisticated and uncertain that it appears most reasonable to adopt either the point of view established by the majority (or convention) or to treat them not as a matter of science but of belief. Into the latter category fall two questions that are possibly some of the oldest ones: "Where does randomness come from?" and "How does order arise?" Everybody who has pondered about the basic principles lying at the foundations of Nature must have encountered these questions. From daily life as well as from traditional learning we have grown accustomed to the almost obvious perception that complicated, irregular, imbroglio behaviour is possible only in very complicated systems. Examples are the vast number of molecules in a balloon filled with gas or a crowd of excited football fans just after the unexpected announcement that *the* football game of the season is cancelled. In the case of such complicated systems we are usually not able to unravel for the individual events unambiguous connections between cause and consequences, that is, we are not in the position to predict the detailed behaviour of the system and thus consider it to be ruled by chance.

It is true that scientists always nourished the hope that in principle such an interpretation in terms of random behaviour and our inability to predict the future of the system could be removed once we had a more complete knowledge of its details. For a long time the following point of view was accepted: if we obtain more precise knowledge on the details of the interaction of all elements of a complicated system and collect more detailed information on their initial conditions then we will be able to predict its behaviour over long periods and the randomness "will become smaller and smaller" as we increase our knowledge.

Similarly natural appears the idea that there must be an "organizing beginning" or "creator" behind each sophisticatedly organized structure which exists in a stable form or which emerged out of a background of noise and disorder. It is precisely for these reasons that the strongly regular structures of clouds or the hexagons of volcanic origin depicted in Fig. 1 appear so imbroglio and even mystic.

In our short excursion to follow below we try to get a glimpse of the beauty of the most intriguing facets of modern nonlinear dynam-

Fig. 1. Hexagonal volcanic structures (Kuril island "Kunashir")

ics, *disorder* versus *ordered structures*. Chaos and order do not just coexist in Nature, they arise due to the same general principles, laws, or conditions which justifies to discuss them in parallel.

In the last few decades two distinct discoveries have been made that have completely changed our perception of the nature of randomness and regularity. It has turned out that very simple systems are able to exhibit random behaviour. These are systems evolving or living according to very simple rules (or having a small number of elements or degrees of freedom). Randomness is an intrinsic property of such a system and is not imposed on it by its surroundings or through external forces. It is impossible to remove it through more detailed studies of the system. Such a randomness exhibited even by simple regular systems is what we refer to as *dynamical chaos*.

The other discovery is related to the awareness and experimental confirmation of the fact that from initial disorder simple as well as complicated highly organized structures may spontaneously arise, continuously developing and evolving. This process is referred to as *self-organization*.

Consider, for example, a group of workers. We then speak of organization or, more exactly, of organized behavior if each worker acts in a well-defined way on given external orders, i.e., by the boss. It is understood that the thus-regulated behavior results in a joint action ...

H. Haken
"Synergetics"
(Springer, Berlin, Heidelberg, New York 1978) p. 191

6.2 Randomness Born out of Nonrandomness

Let us start with chaos. The main difficulty which we have to overcome is a psychological one. We are used to the notion that simple systems (like a swing, a marble in a chute, etc.) display a very simple behaviour. Knowing the rules such a simple system obeys and its initial conditions, we may predict its behaviour over periods of time of any desired length. For example, using Newton's laws we can list all future solar eclipses not only for the next few hundred but even

Self-organization:
Long-living systems slave short-living systems.

H. Haken
ibid. p. 191

thousand years! We need relatively little time to evaluate them, hence, this is indeed a real prediction.

What are then the sources within similarly simple systems – systems living according to simple rules – which give rise to unpredictability, irregularity, randomness? To develop an understanding of such peculiar features, let us start with a simple example.

Presumably all of us encountered back in our schooldays sequences of numbers like the arithmetic series $x_{n+1} = x_n + a$ and the geometrical series $x_{n+1} = bx_n$. Let us refer to these series as systems. In this interpretation the label or number of each member or element may be viewed as a time (measured in discrete units) and x_n as the state of the system at the moment n. The behaviour of such simple systems is indeed very simple, because their states are completely predictable for as long as you like. For arbitrarily large n we have $x_{n+1} = x_1 + na$ and $x_{n+1} = b^n x_1$, respectively. Thus a knowledge of the systems at $n \to \infty$ does not require that we explicitly know its states at all intermediate moments of time in their "life". It is sufficient to evaluate x_{n+1}. Is that always true?

Let us consider a series which is almost equally simple and defined by the rule $x_{n+1} = \{2x_n\}$ where the symbol $\{\ldots\}$ stands for taking only the fractional part of the number within these brackets. Obviously the x_n represent points in the interval (0,1) from 0 to 1 of the number axis. As an example we take for x_1 the number $1/5$. Whence follow $x_2 = 2/5$, $x_3 = 4/5$, $x_4 = 3/5$, $x_5 = 1/5 = x_1$. Indeed, everything is simple: we observe a purely regular motion and for $x_1 = 1/5$ we are able to predict the state of the series after arbitrarily long periods of "time", that is for large n.

Let us give it another try. As the initial number we now take the fractional part of π. With a pocket calculator you can easily verify that with this we get $x_1 = \{\pi - 3\}$ and a chain x_2 , x_3 , ..., x_n which never closes and that the resulting sequence of numbers is similar to one of random numbers. We are no longer able to predict the state of the system. Not even for a few steps ahead. It is now impossible to skip intermediate steps without losing the ability to predict the "future". To give the explicit value of x_{n+1}, we have to evaluate all x_n up to the $(n+1)$-th step. Since the computational time is of the same order as the time of the moment the prediction is made for, that implies unpredictability (chaos) which is characteristic for such a random series. The same will hold for any series if its first element x_1 is an irrational number. And the overwhelming majority of numbers in the interval from 0 to 1 is of this type.

We would call the same process as being self-organized if there are no external orders given but the workers act together by some mutual understanding, each one doing his job so as to give rise to a joint action ...

H. Haken
ibid. p. 191

As illustrated in Fig. 2, the motion of our simple system may also be represented graphically. The two axes represent x_{n+1} and x_n, respectively. The area in between the lines $x_{n+1} = 2x_n$ and $x_{n+1} = 1{+}2x_n$ within the interval $(0, 1)$ is the graphical representation of the rule: extend the initial bit by two and take away the integer. Having reflected the trajectory off the bisectrix it converts the final result of the preceding step into the inital value of the following. The closed trajectories (bold lines) correspond to periodic motions of the system. The tra-

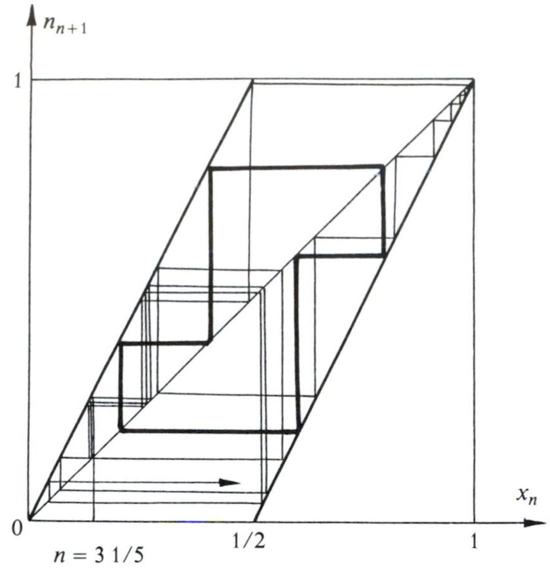

Fig.2. Periodic (*bold lines*) and random (*thin lines*) trajectories in the simple system defined by $x_{n+1} = \{2x_n\}$

jectories (thin lines) which never close are examples of complicated, chaotic motions.

For computer calculations it is convenient to represent x_n in a binary system. In that case each subsequent value in the series is obtained from the preceding one simply by moving the number one digit to the left (which corresponds to multiplication by two) and by suppressing its integer part. For example, if $x_1 = 0.101100010111\ldots$ then we get $x_2 = 0.01100010111\ldots$, $x_3 = 0.1100010111\ldots$ etc. Periodic sequences correspond to rational numbers and random series to irrational numbers. (Let us recall that the irrational numbers are in the majority, implying that almost all arithmetic sequences should be random; usually we do not pay attention to this point because we use normally rational numbers.)

Brownian motion – the drunken man's walk.

From these examples we have seen that simple fully deterministic systems may indeed give birth to chaos.

6.3 An Unstable Path and Steady Motion Are They Incompatible?

The points x_n, $x_{n+1}, \ldots$ wander in a random manner within the interval (0,1) extending from 0 to 1. Yet, this chaos does have its own order. To see that in a more pictorial way, let us go over from a series of real numbers to one with complex numbers. As an example we take the series represented by $x_{n+1} = x_n^2 + C$. The majority of the resulting points will no longer be found within the interval (0,1) but in the plane defined by the real and imaginary parts of x_n namely $\mathrm{Re}\,x_n$ and $\mathrm{Im}\,x_n$, respectively. For $n \sim 10^6$ and $C = 0.32 + i0.43$; $\mathrm{Re}\,x = (-2.0\ldots+2.0)$; $\mathrm{Im}\,x = (-1.8\ldots+1.8)$ this leads to the picture shown on Plate 1.

Egg or chicken? Who was first?

Each one of the points of this series with the coordinates Re x_1, Re x_2, ..., Re x_n, ... and Im x_1, Im x_2, ..., Im x_n, ... is random. Yet, it is problematic to refer to this manifold as being random. For example, it resembles only vaguely the random figure formed when a handful of sandgrains is thrown at random to the ground. The surprising beauty of this figure is, as a matter of fact, a consequence of the point that the random series emerged in the given case from a deterministic system. It has its own order, that is the dynamical chaos is in a specific way "organized". This is a remarkable phenomenon which demands a more detailed discussion. However, first we have to answer the question left out earlier: amongst the manifold of series' the majority of which are random, there are also periodic ones; maybe they are the only ones that are indeed observed?

No. They are unstable, implying that for arbitrarily small uncertainties or variations in the given x_1 we will already have a completely different series, see Fig. 2. Here we note a manifestation of another feature of dynamical chaos, an extremely strong sensitivity to changes in the initial conditions. Thus, in spite of the fact that there are many periodic series, even infinitely many (yet, there are nevertheless many, many more nonperiodic ones) it is practically impossible to observe them. However, the nonperiodic trajectories are also unstable. Why is it then that they are observed at all? The point is that there is an almost continuous manifold of them and that they are indistinguishable. Indeed, let us assume that because of a small perturbation (or uncertainty in the given x_1) we obtained not the "correct" series, nevertheless the other one which we obtained belongs also to the same manifold of unstable trajectories. Thus we will always observe one or the other of them. Consequently we are led to a remarkable conclusion: although the individual trajectories of a stochastic manifold are unstable and thus not observable (not realized), the manifold itself is stable and at least one of its many trajectories (covering almost the entire manifold) is observed!

6.4 Does Chance Rule the World?

Why didn't we realize earlier the random behaviour of nonrandom systems? Maybe, we were only picking artificial examples? Maybe life is different and real systems with chaos are described by different equations? For example, could it be that Newton's equations guarantee only a regular behaviour of mechanical systems?

Two combined periodic motions: a swing and its driving force.

Of what type is the combined motion?

Let us consider a very simple system, a swing. Of course, it is described by the simple equations of mechanics which some of us encountered at school. But apparently the example is not a good one: as we know, the swing oscillates periodically; however, its period is also determined by the actions of the person using it, standing up or squatting. The effective length of the swing may thus be altered to accelerate it or (by getting up out of phase) to slow it down. Let us now deprive the person on the swing of his will power. Let us assume that it is now a robot who squats and stands up strictly periodically! What will happen? – The swing will oscillate completely randomwise, without any random forces acting on it! Squatting strictly periodically the robot will out of phase "lengthen" or "shorten" the swing and consequently add energy to the swing (or pendulum) or take it away, respectively. The observed randomness is due to the fact that the swing moves with different speeds depending on its angle of deviation from the vertical. For example, close to the position "head over heels" it slows down drastically and at the turning points it even comes to a stop. Thus we see that even real mechanical systems that obey Newton's laws, behave chaotically. That is, they generate randomness! Why didn't we observe it earlier? It seems appropriate to respond: "We saw it, but we did not realize what we were seeing." The traditional ways of thinking did not allow us to take serious individual experiments demonstrating chaos in simple systems (and therefore not fitting into the framework of established theories). In some way or the other they were "explained" in terms of natural fluctuations or by the salutary influence of noise that could not be accounted for in the calculations.

The result of these two strictly periodic motions is chaos!

The sensitivity to the initial conditions typical for chaos is due to the instability of the motion. In a very pictorial and direct way reflected by historical developments.

History would have a very mystic character, if randomness played no role in its evolution. However, during different periods of time in the evolution of society and in different countries this randomness has manifested itself differently. In periods of *stable* evolution, randomness (like the death or murder of a political leader, natural calamity, etc.) pushed the evolutionary path of the society only from one trajectory to another one lying in its immediate neighbourhood: the welfare of the people changed a bit, building plans were altered, interior or exterior politics came to be harsher or more moderate. A qualitatively different picture is observed in a period of *unstable* evolution (just before the outbreaks of wars or revolutions, in periods of disorder and unrest): small random deviations have led to completely different routes in the subsequent development of the society. In such periods it appears indeed as if "chance rules the world". But in such a case one usually forgets that such significant results due to the action of randomness are in reality only possible because the "flow of life" which was completely changed by an insignificant randomness was unstable and randomness just played the role of a trigger.

The reasonable man adapts himself to the world; the unreasonable one persists in trying to adapt the world to himself. Therefore all progress depends on the unreasonable man.

G.B. Shaw
Man and Superman, III

The following children's rhyme illustrates such a situation neatly:

For the want of a nail, the shoe was cast.
For the want of a shoe, the horse was lame.
For the want of a horse, the message was late.
For the want of a warning, the city was lost.

This is a pictorial example for a motion that is at every point unstable: if there were a nail, then the horseshoe would have been fixed; if the horseshoe were fixed then the horse would not have been lame; if the horse were not lame (or if there were a spare one) then the message would have been in time; if the message were received in time then the city would not have been lost.

6.5 What is the Character of Nature? Integer or Fractal?

Plates 1,2
Fractal manifold

A glance at Plate 1 is sufficient to convince us that the manifold of points depicted in the plane of random sequences generated by a deterministic system is remarkably regular. Similar manifolds with microstructures are depicted on Plates 2–7. According to the words of Ruelle they possess an "aesthetic attractiveness". These systems of curves, these clouds of points resemble sometimes fireworks of galaxies, sometimes strange mysterious brushwood. This is a sphere of research in which new harmonies will be discovered. Such manifolds with "floating" rarefying structures (which have as a rule noninteger dimensions as will be explained below) are called *fractals*.

Plate 3
Fractal "spiral street"

An example of fractals is given by the sophisticated evolution of the discrete analogue or predator-prey system modelled by the Lotka-Volterra equations

$$\begin{pmatrix} x_{n+1} \\ y_{n+1} \end{pmatrix} = \begin{pmatrix} x_n \\ y_n \end{pmatrix} + \frac{h}{2} \times \begin{pmatrix} f(x_n, y_n) + \varrho f[x_n + \varrho f(x_n, y_n) y_n + \varrho g(x_n, y_n)] \\ g(x_n, y_n) + g[x_n + \varrho f(x_n, y_n) + \varrho g(x_n, y_n)] \end{pmatrix}$$

Plates 4–5
Predator-prey system

Plate 6
Predator-prey system

with $f(x, y) = \alpha x - \beta x, y$ and $g(x, y) = -\gamma y + \delta x y$. Plates 4–5 show its evolution for $h = \varrho = 0.739$ and Plate 6 for $h = 0.8 = \varrho = 0.8$.

Fractals arise also in the stochastic manifold of the nonautonomous self-oscillations of the system

$$\ddot{x} + \gamma \dot{x} + x + x^3 = F_0 \sin(\omega t)$$

Plate 7
Nonautonomous self-excited oscillations

depicted on Plate 7 for $\gamma = 0.1$ and $\omega = 0.95$ and the following values of the amplitudes F_0 of the external force:

a) $F_0 = 8$, b) $F_0 = 18$, c) $F_0 = 20$, d) $F_0 = 25$.

Although the term fractal was rather recently introduced into science by Benoit Mandelbrot, fractal objects have been the subject of research for quite a long time. Judging from their huge number, Nature "likes" fractal forms very much. Amongst them we have colloids, electrical discharges, porous solid states, coastal lines, the structures of turbulent flows (the majority of whose trajectories are points generated by deterministic systems with chaotic behaviour) and many more. No doubt, the beauty of fractals is linked to their property of self-similarity which manifests itself in repeated patterns at changing scales of the figures. This is nicely illustrated on Plate 8 taken (together with Plates

Plate 8
The infinite complexity

1–6) from a book by *Peitgen* and *Richter* with the title *The Beauty of Fractals*.

Plate 9
Nonperiodic lattice evolution

The symmetric lace pattern on Plate 9 as generated by a dynamic system exhibits a surprising beauty and charm. However, strictly speaking this illustration is not fractal – it is rather like a mosaic consisting of a large number of elements of finite (!) size. Thus it does not allow for an infinite detailing which is an inherent feature of true fractals.

The equations leading to Plate 9 model a lattice in terms of the equations

$$\frac{du_{jl}}{dt} = u_{jl} - (1 + i\beta)|u_{jl}|^2 u_{jl} + \kappa(1 - ic)(u_{j,l+1} + u_{j,l-1} + u_{j+1,l} + u_{j-1,l} - 4u_{jl})$$

with $j, l = 1, 2, ..., N$ and $N \gg 1$. For a large supercriticality, that is for approximately $r = 1/\kappa \geq 0$, the discrete lattice described by these relations gives rise to spatio-temporal chaos. In this process the short-wave excitations play the most important role. In the extreme case the system leads to so-called π-π-oscillations corresponding to anti-phase oscillations of neighbouring lattice elements. The arising chaotic structure is very fascinating – it is a kaleidoscope of pictures that never repeat in time. The complexity of its spatial structure is determined by the number of elements in the lattice, see also Fig. 3 (to which we will come back below when discussing Plates 30–32).

If we magnify under a microscope an arbitrarily small area of a fractal (or stochastic) manifold, i.e. if we increase the resolution significantly, then we observe again an exceptionally complicated picture with a manifold of details. This is nicely illustrated on Plates 2,3 and 8. And thus we may proceed to infinity! – a most extraordinary fact. Really up to infinity? What determines the threshold for this chain of subsequent reductions and detailizations? Before answering this question, let us consider what could possibly perturb such a process. Let us recall the underlying rule: what we need is that in between two arbitrary trajectories (for example, strange attractors), we can find as

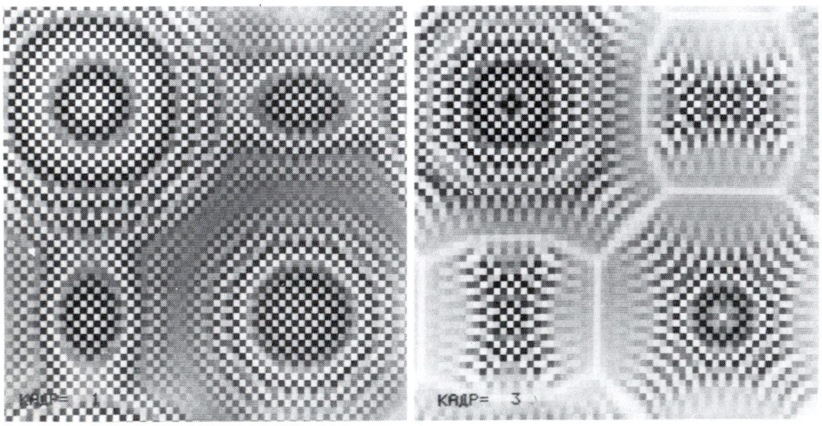

Fig. 3. Modulation waves and π-π-oscillations in the lattice-model referred to in the discussion of Plate 9

many further trajectories as we like (!) – no matter how close our initial two trajectories are to each other. The same holds for the paths of marked particles which by a pulsating stream are mingled in real (Euclidean) space, see Plates 10 and 11. They show the chaotic mixing of an admixture of laminar (regular) flows. The fluid rotates between two cylinders with different rotation axes, the rotational speed of the inner cylinder is modulated with constant frequency. When are we allowed to consider these trajectories – even if we take them to be infinitely thin – as being different? Obviously only if the space in which they move is continuous!

Plate 10, 11
Chaotic mixing of laminar flows

It is the discreteness of space (if it corresponds to the given problem) that is the cause for limitations in subsequent detailizations of fractal structures. And for dynamical chaos it is the cause of the existence of chaos itself. Let us look into this in more detail.

Let us imagine space to be discrete (no matter whether it is the real space in which liquid particles move or the state space or phase space). This discreteness may be caused by different reasons. In the case of the flow of a liquid this is naturally related to the impossibility of considering the path of a fluid particle to be an infinitely small line. In that case the trajectory of each particle will be closed and chaos dissapears.

The discreteness of space may also be determined by deeper reasons. In the first place, it may be related to the discreteness of the elements which make up the flow of liquid or gas (for example, the molecules) in real space. It may also be related to the quantum nature of the objects under investigation because in that case the quantization condition gives rise to a discrete phase space.

6.6 Fractal Fingers

In passing, we mentioned that fractal manifolds have noninteger dimension. What does that mean? Maybe, noninteger dimensions are the result of approximations? It is just the opposite. Estimating the dimensions of an object by eye we make them due to our habits integer ones. Indeed, what is the dimension of an entangled "ball" or clot of thin thread? If we look at it from the far distance we notice a point object implying zero dimension. If we come closer then its dimension is three. If we want to consider this object in detail and follow the individual bits of thread, the tiny building units which make up this clot then its dimension seems to be unity. But why does the dimension change in jumps? Well, because this is what our perception is used to and, consequently, it *appears* to us to be true. To obtain the correct answer in the analysis of such unusual objects like fractals (to which, in a crude way, our ball or clot also belongs) we have to define the term dimension more rigidly.

The great desideratum for any science is its reduction to the smallest number of dominating principles.

J.C. Maxwell

This may be accomplished by counting the number of elements of the object of interest found within a sphere of radius r with its center within the object under consideration, see Fig. 4. The number N of elements found within such a sphere will be proportional to the D-th power of the radius of the sphere with D standing for the dimension of the system. That is, using unity for the associated dimensional constant we have $N = r^D$ or $D = \ln N / \ln r$. For an object in the form of a straight line the number of its elements within such a sphere increases by a factor of three if the radius is tripled, that is $D = 1$. If the object is the close packing of elements in a plane then tripling r implies that N is increased nine times. Consequently we obtain the dimension $D = 2$. So far the rigorous definition gives only the obvious traditionally expected results. However, let us turn to the less trivial situation depicted in Fig. 4. For this structure tripling the radius r of the sphere leads to five times more elements within the sphere, i.e. the number of elements

Physics is experience, arranged in economical order.

Ernst Mach

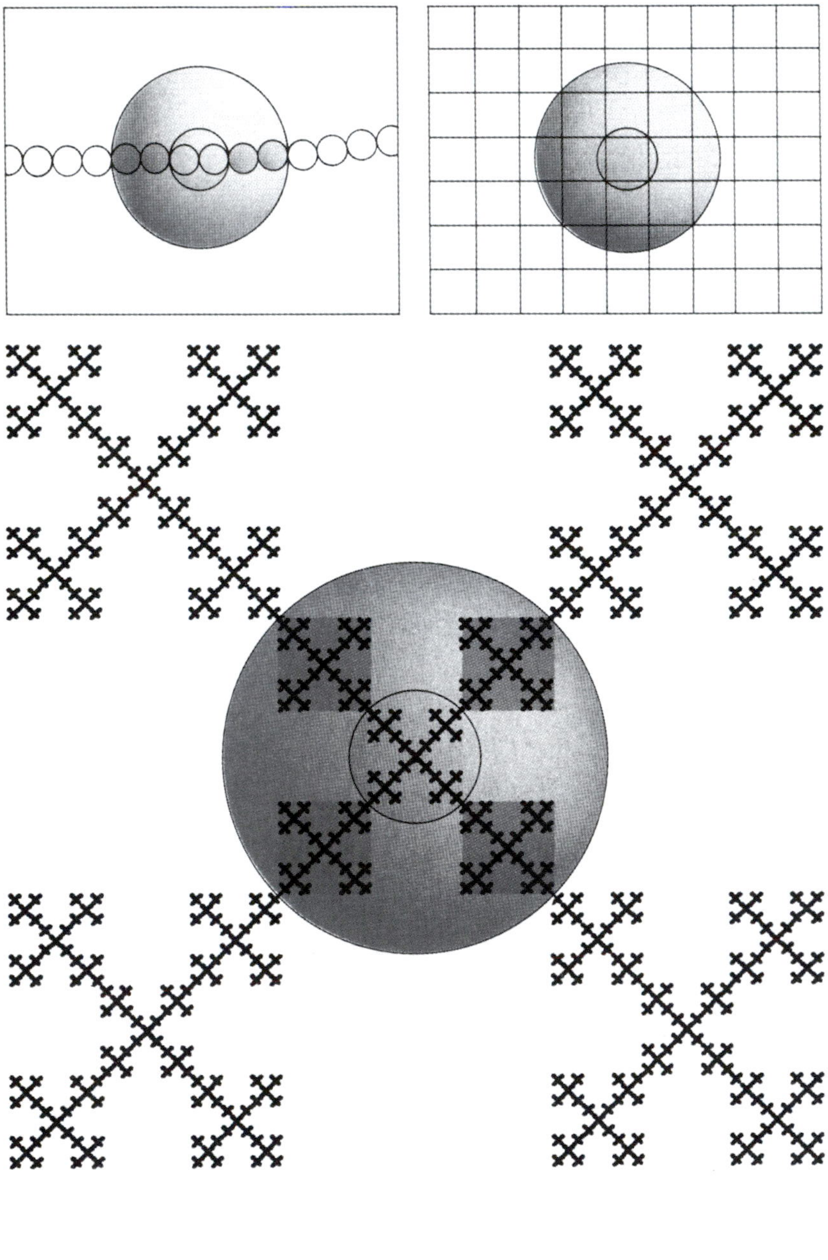

Fig. 4. A fractal "snowflake"

grows faster than for one-dimensional object and slower than for a two-dimensional one. This is related to the fact that our structure is everywhere "porous", "holey" or fractal so that its dimension turns out to be a noninteger one, namely $D = 1.46$.

Plate 12
Viscous fingering

Similar fractals close to the ones of a plane are, for example, obtained for a less viscous liquid penetrating at a high flow rate into a more viscous one, when both are flowing down a thin cuvette, see Plate 12. The Plate shows the patterns observed in a thin gap between two glass plates. The miscible liquid water entering through the opening at the centre is progressing into the radial direction (Hele-Shaw cell) and interacting with glycerine. There is no surface tension between the two fluids, and a diffusion-limited aggregation structure with fine branches appears. We notice the viscous fingers of the water solution of the polymer. The growth of these remarkable structures is connected with the presence of a pressure field which pushes the water into the cuvette. Their filigree configuration is determined by the competition between the growth of neighbouring elements due to the pressure gradient in their vicinity. The threshold limiting the thinness of the fingers is determined by the surface tension. Thus one obtains the most delicate filigree-like picture for liquids with rather similar surface tensions. Structures of the same type are obtained in computer simulations, see Plate 13.

Plate 13
Fractal structure

This illustration displays the fractal structure as obtained in computer simulations modelling diffusion-limited aggregation. The growth of the structure develops with time. Its oldest part is given in white.

6.7 Self-Organizing Structures

This is the right place for a pause to consider the question: why is it in spite of the presence of noise, different forms of inhomogeneities etc. that the area occupied by the water (see Plate 12) is so ordered and extremely beautiful? Why does it not look like a blot with random contours? At this stage we have to go over to the second part of our excursion into the realm of *Nonlinear Physics – Chaos and Order*, to the problem of order arising from disorder, to the problem of self-birth or self-generation of structures.

Plate 14
Spatial field distribution of a nonlinear lattice

Plate 14 illustrates the birth of an ordered structure out of an initially disordered system as observed within the computer simulation of the two-dimensional medium depicted by the lattice model referred to in Plate 9. Out of disorder order is seen to arise – but maybe that happens only in computer experiments? Maybe, the in reality ubiquitous presence of noise and fluctuations does prevent the formation of such a self-organization in Nature?

Plates 15–17
Birth of hexagonal lattice

Let us go over to reality. The sequence of pictures shown on Plates 15–17 illustrates the self-generation of a regular hexagonal lattice out of an initially mixed inhomogeneous liquid. This (Rayleigh-Bénard) lattice is the result of convective flows emerging in a horizontal layer of the liquid (in the given case it is a silicon oil) when uniformly heated from below. In this process the upper surface is free and surface tension plays an important role. Visualization is facilitated by adding aluminium powder to the liquid.

The hexagonal convective lattice arises only when the temperature difference between the heated lower liquid layer and the cold upper surface exceeds a certain critical value. It is only in this case that the molecular heat exchange (in the absence of macroscopic motion in the respective liquid layer) is not able to comply with its "task" of transporting the temperature, thus leading to convective instabilities within the layer. They start off from fluctuations (read "disorder"!) and yield vortex motion with characteristic scales which depend on the thickness of the layer. (This instability is related to the fact that the lighter heated part of the liquid follows Archimedes' law and moves upward forcing the heavier cold parts of the fluid to move downwards.) Thus we have again an instability. But now in the opposite sense: not giving rise to chaos but to order, to regular structures! The instability of a quiet, equilibrium state of the liquid leads to the preferred growth of the motion at individual characteristic scales at the expense of others and thus in the very end to regularly structured flows.

Plates 18–20 provide an excellent visualization of the interaction of a 2-d wake with a jet plume. These photographs show the flow fields that are formed when the wake of a 2-d bluff body interacts with the plume of a slot jet. The annular air flow velocity is fixed at 27.5 cm/sec. For an averaged air jet velocity of 18.5 cm/sec (*left* of Plate 18), the alternating vortex structures shed from the bluff body are clearly formed after about five bluff-body widths downstream. The wake from the bluff body is significantly modified as the jet velocity increases. At a jet velocity of 37 cm/sec (*right* of Plate 19), the shear layer velocities of the jet and the wake are nearly equal. At still higher velocities, the jet begins to dominate as indicated by the change in the direction of rotation of the vortices.

Plates 18–20
Interaction of 2-d wake and jet plume From the left to the right and going over from Plate 18 to 20 the jet velocity is increasing and takes on the following values: 18.5 cm/sec, 22.2 cm/sec, 27.7 cm/sec, 37 cm/sec, 55.5 cm/sec, 63.3 cm/sec

Similarly regular structured flows of liquids are quite often met in Nature. Amongst them are for example large vortices in the atmosphere of the Earth and also the regular structures of "topographical vortices". These topographical vortices of volcanic material were formed as a consequence of the particular profile of the shelf.

But nevertheless, when we talk about the self-organization, self-generation of structures and follow their evolution towards completeness we (possibly subconsciously) have in mind something greater and by far more fundamental and unusual than simply the appearance of regular structures similar to the wave-like ones depicted in Fig. 5 and on Plate 24, say. Figure 5 illustrates the regular three-dimensional structure of waves on the surface of a fluid. Due to instability they arise from a homogeneous train of steep plane Stokes waves. Plate 24 shows the structure due to capillary Faraday ripples on the surface of a horizontal liquid layer in a flat cuvette whose bottom oscillates in space with a frequency of 120 Hz.

Plate 24
Capillary Faraday ripples

Fig. 5. Three-dimensional wavestructures

6.8 Singles

As a matter of fact, we hope that as the result of the evolution of the instability it will be possible to observe the self-generation and the stable existence of individual localized (!) structures which behave in some way like particles and preserve their individuality in their interaction with each other and under the influence of external fields. Such localized structures or "self-structures" do indeed exist. We can most readily observe them in inhomogeneous external fields.

Plate 21
Flow vortex

For random initial conditions Plate 21 depicts the three-dimensional localized vortex self-structure arising above a smooth indent at a surface over which the wind (or a gas flow) blows. The topology of this localized pattern changes only for finite variations of the parameters. In this example the speed of the wind across the indent is the critical parameter of the system.

Plates 22–23
Convective polyhedron

Plates 22 and 23 show the self-generation of a convective self-structure in the form of a polyhedron from arbitrary initial conditions in an inhomogeneously heated layer of silicon oil. Independent of the form of the heater, the polyhedron emerges during the inhomogeneous heating of the layer of silicon oil as a stable being! The transformation of such a localized structure from one form to another one occurs only for finite changes in the thickness of the layer or in the temperature difference.

6.9 The New Life of an Old Problem

Plates 25–28
Dynamical turbulence. The evolution of spatio-temporal chaos – dynamical turbulence – against a background of capillary Faraday ripples

Structures in nonequilibrium media are amazingly stable. For example, for convection or Faraday ripples they remain unchanged even for supercritical numbers (well above the temperature difference at which convection arises) – that is, even for numbers largely exceeding the ones marking the onset of turbulence. Plates 25–28 display the evolution of spatio-temporal chaos – dynamical turbulence. It arises against a background of turbulent Faraday ripples for different vibrational amplitudes of the bottom of the plane cuvette with the liquid on the surface of which the events are taking place. As the amplitude of the vibrations grows (i.e. as the values of the critical parameter increase well beyond the critical one) the regular lattice of quadratic cells becomes more sophisticated (compare with Plate 24). On the background of the initial pattern modulating structures appear. Amongst them there are also localized structures like defects in the periodic lattice, see Plate 28. They are in motion and interact with each other. In the general case their motion is chaotic. It is the chaotic spatio-temporal dynamics of defects in a certain region of supercriticality, which is referred to as *turbulence*.

The interaction laws between defects are to a good degree universal. The topology of defects and their respective transformations turn out to be similar for Faraday ripples and thermal convection. Similar features are displayed by the spiral waves of synchronization performing a random walk on a two-dimensional lattice of generators (described above in the context of Plate 9), see Plate 29. Another example is given by the spirals of liquid crystals depicted in Fig. 3 and on Plates 30–32. They show the life of defects (of the smectic C-type). Most typical defects have the form of four-fingered disclinations. In particular, if one starts to rotate the knots seen in Fig. 3 (a transformation which does not affect their topology), then one obtains spirals similar to those observed in two-dimensional chemical reactors, where autocatalytic Belousov-Zhabotinsky reactions take place, see Plates 33–35. If there are many of these spirals, their interaction which follows dynamical (regular) laws, gives rise to the spatio-temporal chaos or "chemical" turbulence exhibited on Plates 36 and 37. The change in colour on Plates 33–37 reflects the changes in the relative concentration of the interacting reagents.

Plate 29
Localized defects in the Ginzburg-Landau model (*left*) and defects on a two-dimensional lattice of generators (*right*)

Plates 30–32
Life of defects in liquid crystals

Plates 33–35
Spiral waves

Plates 36–37
"Chemical" turbulence

6.10 Spatial Evolution of Disorder

The notion of turbulence as a spatio-temporal chaos of structures is well-known to all of us. You will certainly agree when you look at the two photographs on Plates 38, 39.

Plate 38
Wave instability on vortices

The concept is also applicable to anisotropic media and in particular to shear flows. Structures arise as the result of evolving shear instabilities. Their dynamics and the character of their interaction along the flow becomes more complicated and as a result one observes eventually (downstream) the birth of turbulence.

Plate 39
Birth of turbulence

Turbulence in shear flows may arise along the flow and not just gradually but all of a sudden as demonstrated on Plate 39. It is generated as the result of the evolution of an instability flooded by an axially symmetric stream. In the experiment shown, a laminar flow of air is ejected from a circular pipe with a Reynolds number of around

10000. It is made visible by the aid of smoke. In the exterior region of the stream axially symmetric oscillations arise (due to the instability), then the flow in this region of space curls up into vortex rings. Later on the system goes (along the flow) in no time over to a state of turbulence.

Plates 40–42
Turbulent streams

In our real three-dimensional space the liquid particles flowing initially in parallel, get into each other's way and lose their individual paths, thus giving rise to the instability of the motion of the liquid with the very complicated fractal structures depicted on Plates 40–42. To arrive at such a nice visualization some additives were given into the liquid and laser light was used to induce fluorescence in the flowing liquid.

6.11 What Does Your Camera See When It is Watching TV?

To determine the general laws of the nonlinear dynamics of such structures it is convenient to use artificial nonequilibrium systems that can be prepared in the laboratory or even at home. They allow for detailed studies and a good understanding of the mentioned universality of the topology of these structures and their respective transformations in interactions with each other. One of the possible variants of such devices or artificial "media" can be set up at home with the help of an amateur video camera and a TV set: direct your camera towards the screen of your TV set and transmit the obtained signal of the camera into the video input of your TV set. This "camera + TV + feedback" system is self-excited for sufficiently strong amplification. Naturally, the amplification coefficient has to exceed the characteristic critical value not just temporally but also spatially (which you may achieve by coming closer to the screen and accordingly stretching the picture).

As the result of such self-excitations you observe video pictures on the screen of the TV. In the direct sense they have nothing in common with the pattern "seen" by your video camera. If we experiment in a dark room then we find out that this system is characterized by "hard" excitation. In the dark it is necessary to generate a light pulse to "ignite" the screen. You can watch how the light of the flame of a lighted match brought within the aperture of the objective spreads out across the screen (according to the laws of nonlinear diffusion). The pictures which we thus obtain resemble the ones of the fractal structures looked at earlier. We observe spirals, "viscous fingers" and many others, see Plates 43–46 depicting the video structure of the "camera + TV + feedback" system on the screen of our monitor. Plate 47 shows the spiral structures observed on the TV screen when the video-camera is shifted a little bit around the horizontal axis.

Plates 43–46
"Video-structures"

Plate 47
Spiral structures on TV

The dynamics of the structures in our home experiment with unstable media are of an amazing variety. Depending on the particular conditions, they may either be chaotic or regular. With the help of a separate camera directed towards the screen we obtained Plates 48–52 showing self-oscillations of our "camera + TV + feedback" system: The video structures replace each other periodically as time progresses.

Plates 48–52
Self-oscillations on TV

6.12 Multistability and Memory

We may also use the "camera + TV + feedback" system for completely unexpected purposes, say, for obtaining parallel backward information. Let us ponder a bit about this point.

As many as three decades ago it was clarified that a good number of the processes associated with the vision of a frog (including practically all computations related to the recognition of simple objects like insects) takes place in the retina of its eye. In the language of our computerized world the retina is just an ensemble of parallel processors which spend almost no time in communication with each other. Why are they not communicating with each other? – In the case of a network of only twelve communicating processors almost *all* time is needed for the exchange of information so that there is essentially no time left for the actual evaluation of the information, that is for the recognition of the object.

The simplest architecture of a computer with parallel information processing of pictures is represented by a rectangular matrix of processors, each communicating only with its neighbour. Onto such a matrix we may by the aid of a video camera project the picture which is to be evaluated. Devices of this type are called *Cellular Logic Image Processors* or *Automata*. If we assume that each element of the matrix can only perform a small number of operations, then the number of states which it can receive as a function of the amplitude of the incident signal is also small. Let us try to employ as such a *Cellular Processor* the screen of the colour TV set in our TV-system with feedback. By the aid of a lattice of "open boxes" arranged in front of the screen we divide its area into small cells (the dimensions of which should naturally not be smaller than the ones of the smallest points our TV set can reproduce stably). Each of the cells can now be in one of four independent states (the nonexcited ground state versus three excited states represented by red, blue and green). Thus the number of stably existing pictures (attractors) in our "camera + TV + feedback" system may be sufficiently large. The system may be multistable. The stability of the pictures allows us on the one hand to remember them, on the other hand to distinguish different ones. Plates 53–56 illustrate the phenomenon of multistability for our TV-system. To obtain these pictures we used another rather simple method to split up the screen into small portions: spectacles. As a function of the initial conditions (or depending on perturbations like the hand seen on Plate 53) the system exhibits several stable states.

Plates 53–56
Multistability of our TV

6.13 Nonlinear Dynamics in Society

We would like to add that models and methods of nonlinear dynamics themselves are also continuously developing. Let us mention just one comparatively new direction, the modelling of nonlinear dynamics of extended systems with the help of cellular automata. In this case the "medium" is treated as a set of discrete cells with given rules for their interaction with each other and evolving in discrete time. Such models turned out to be extremely effective for traditional problems of physics as well as, in particular, for problems related to the modelling of evolution processes and of self-teaching (neural networks).

Already for extremely simple interaction laws between each other, the life of cellular automata may turn out to be highly sophisticated and even chaotic, see Plates 57–62. After the experience gained so far with such nonlinear systems we should no longer be surprised by this. If we follow the time evolution of complicated non-repeated spatial structures in cellular automata we will see that neighbouring cells excite each other while the ones which are not so close to each other slow the process down. The colour indicates the intensity of the excitation; black corresponds to no excitation at all.

Plates 57–62
Cellular automata in action

In spite of our bad experience with predictions we dare to assert that within the coming decade the methods of nonlinear physics and nonlinear dynamics will come into vogue not only for physicians and ecologists but also for economists, sociologists and geographers.

Indeed, as clarified most recently, the growth of cities and the evolution of the network of urban and intercity transport resemble strongly the growth of fractal clusters in models with "limited" diffusion.

The majority of specialists concur nowadays that the idea of dynamic equilibrium which lies at the foundation of traditional economical models is not satisfactory. It is realized that it will inevitably be necessary to build nonlinear dynamic models. One approach [as proposed by *Bak, Tang and Wiesenfeld* in *Phys. Rev. Lett.* **59**, 381 (1987)] is named *Self-Organized Criticality*. When the modelled system reaches a critical state, it experiences a self-reorganization and acquires a new form which may be predicted with a sufficient degree of probability. A similar type of prediction – admittedly applied to the past! – was for example made with the aid of the method of critical dynamics in models of arms races between countries which subsequently entered into open war with each other (Alvin Superstein in *Nature* 309 (1984) p. 303).

The methods of critical dynamics are indeed universal. This universality manifests itself also in the character of the changes observed in the dynamics of completely different processes due to changes in the parameters determining the evolution of these processes. Thus it is, e.g. that the same changes (bifurcations) are experienced within the mentioned arms races and the oscillations in radio generators. In Figs. 6a–c (*from top to bottom*) the bifurcation (period) doubling and the transition to chaos as observed in experiments with microwave generators is illustrated.

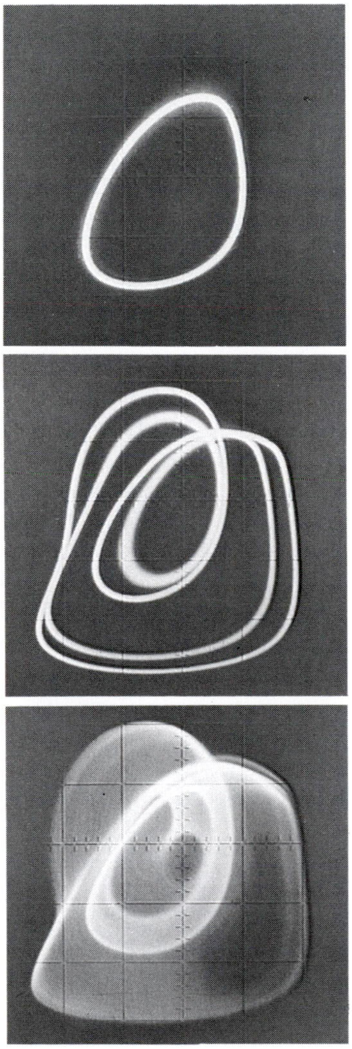

Fig. 6. Period doubling

To close this excursion we return to the lattice model already known from the discussions of Plate 9 and Fig. 3 and display on Plate 63a–d the spatial field distribution of this model (for specific parameters, i.e. $\beta = c = 1.71$, $N = 64$). As in the cases looked at before, the resulting pictures do not just provide valuable information for the specialist, but a rather unexpected beauty making them also attractive to the layperson.

Plate 63a–d
Spatial field distribution in the lattice model discussed in the context of Plate 9 and Fig. 3:
a, b) the dynamics of large "vortices" and the emergence of "spirals" ($\kappa = 16$) *on the left from top to bottom*
c, d) large "vortices" with an excited background on short-wave lattices (π-π-oscillations with $\kappa = 1/16 + 10^{-4}$) *on the right from top to bottom*

Color Plates

The illustrations on the following pages are discussed in Chap. 6 where they are described in a general context. For further details, explanations and the captions to these Plates please consult Chap. 6.

Quite a few of these beautiful photographs could only be included thanks to the courtesy of the colleagues who obtained them and gave us permission to use them in this book. At this juncture we would like to thank all of them again for their help and kind support in enabling us to present such an impressive collection of extraordinary examples showing the beauty of nonlinear science.

1

2

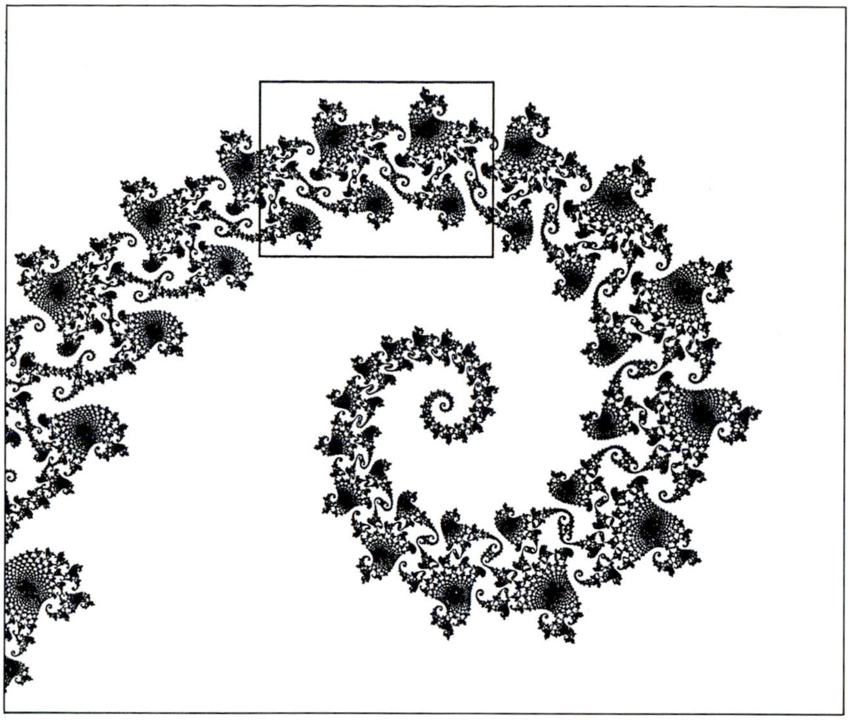

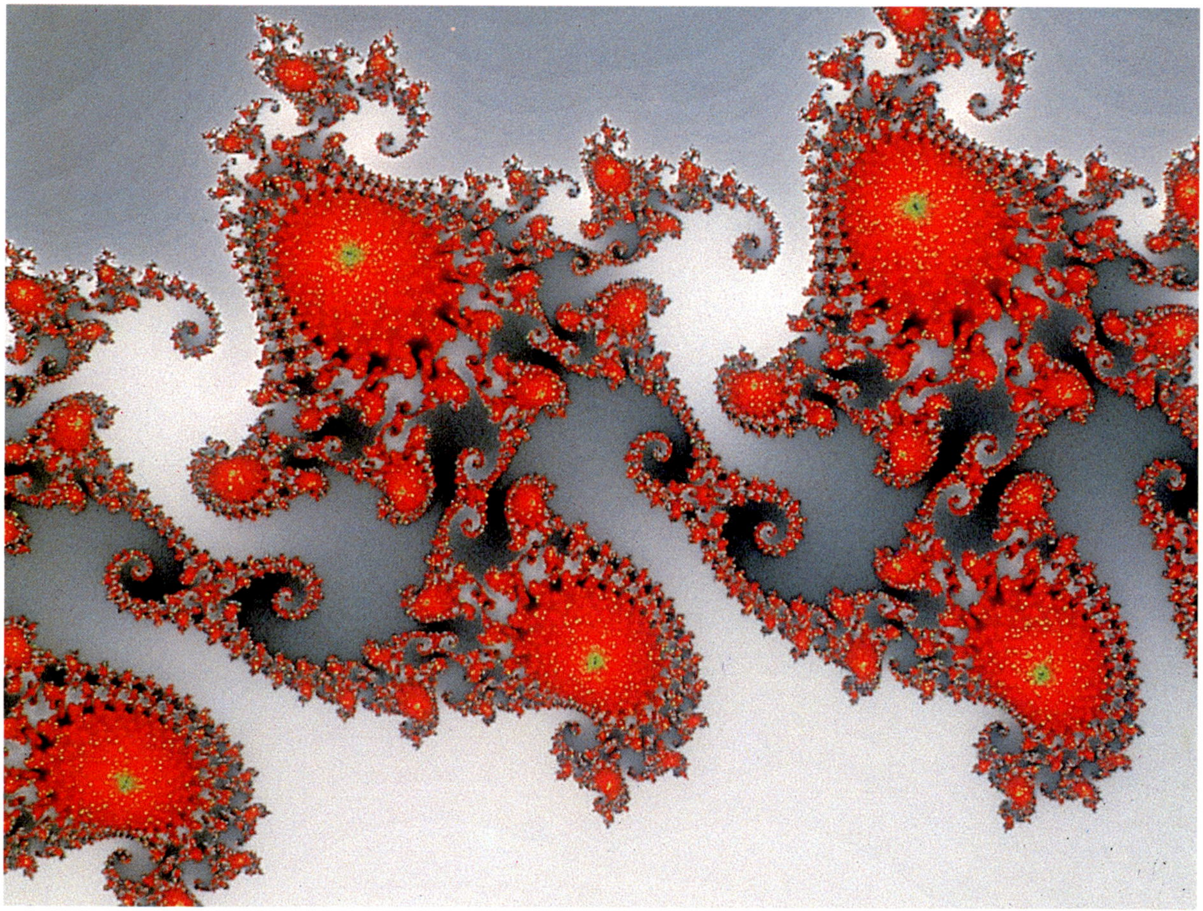

3

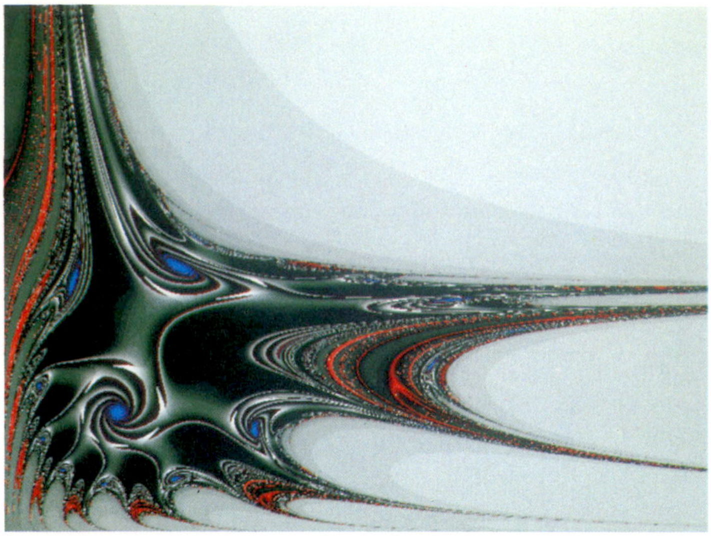

4

5

6

7

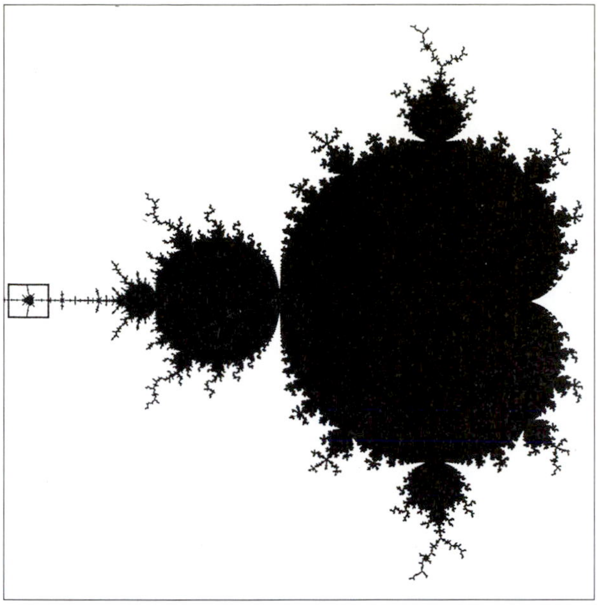

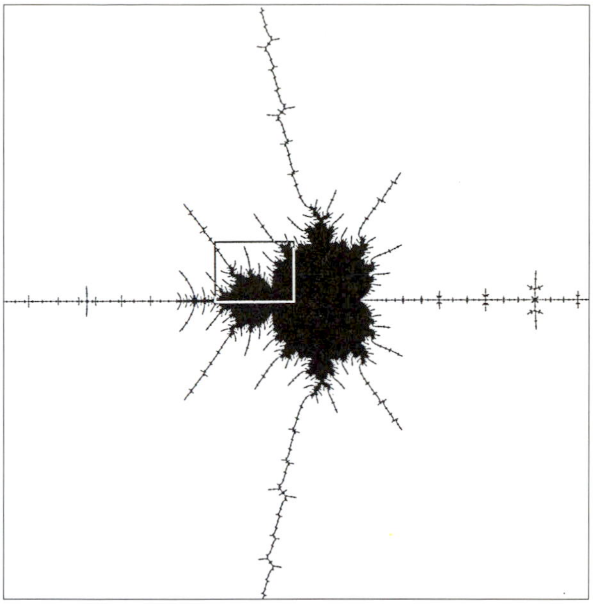

8

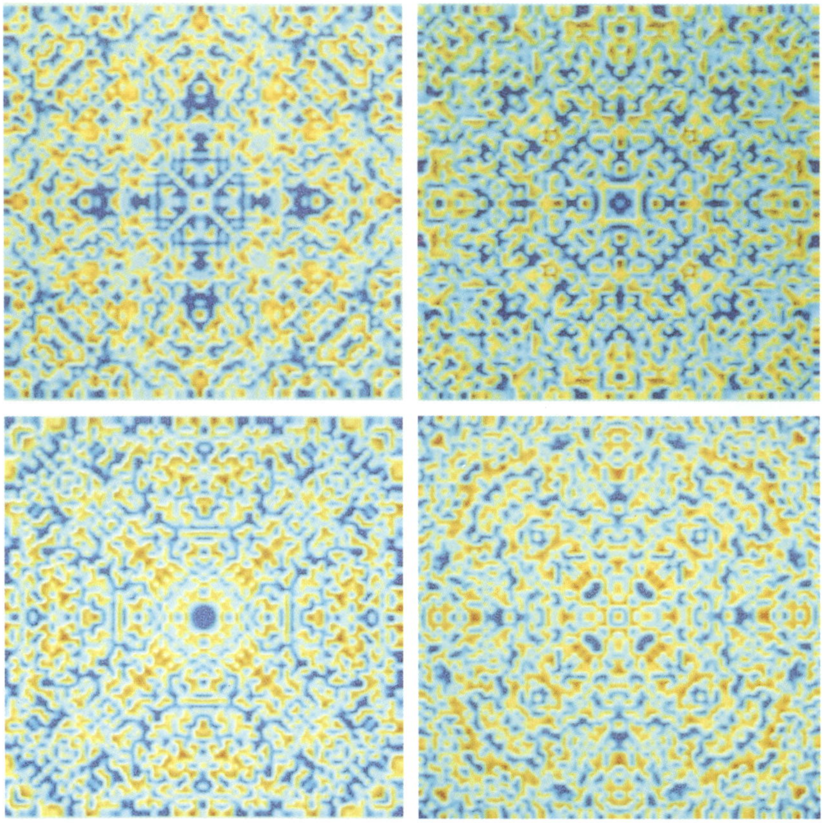

9

10

11

12

13

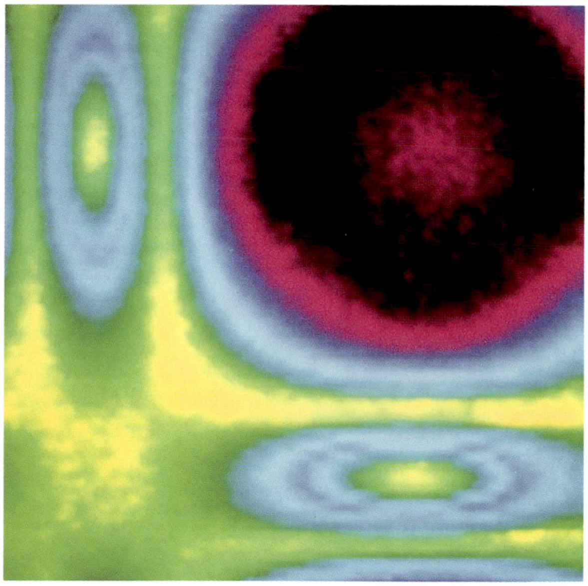

14

15

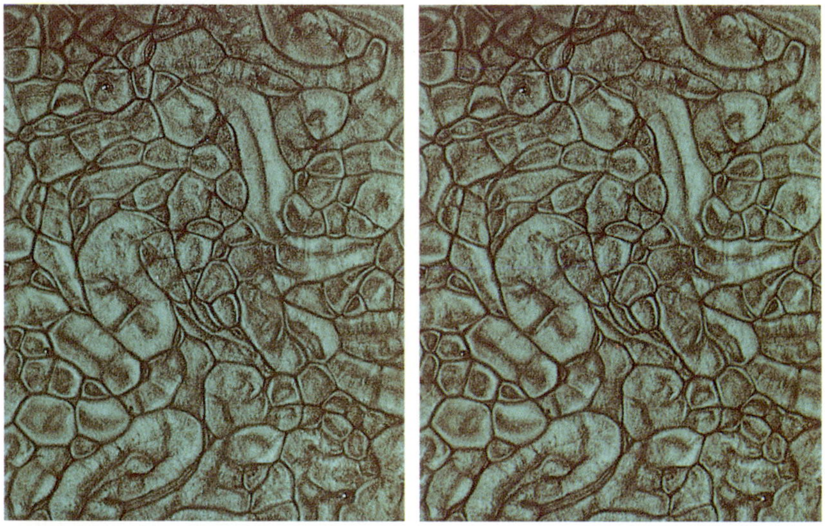

16

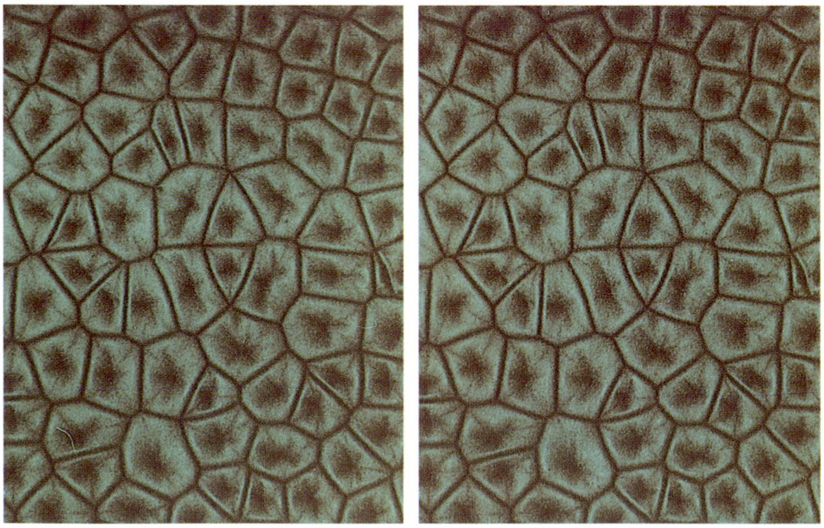

17

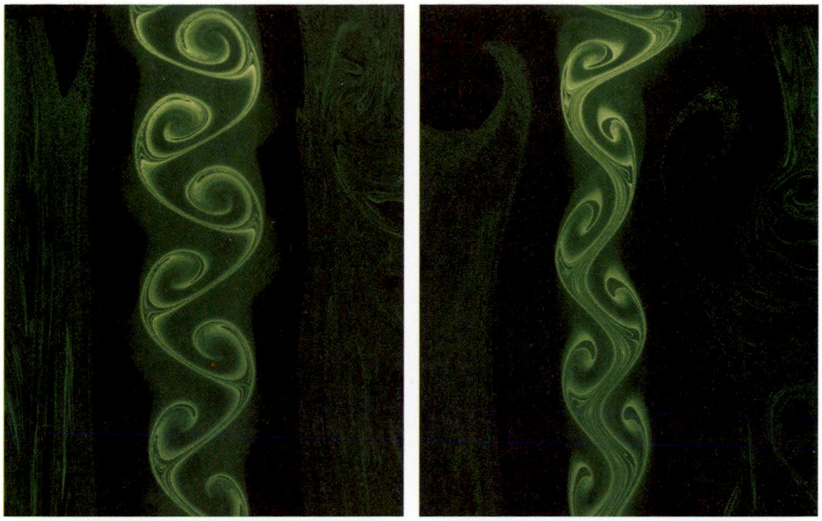

18

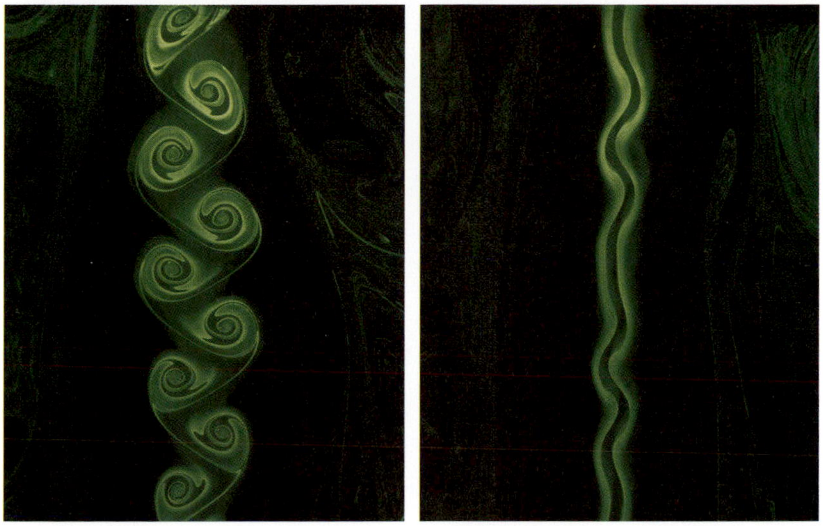

19

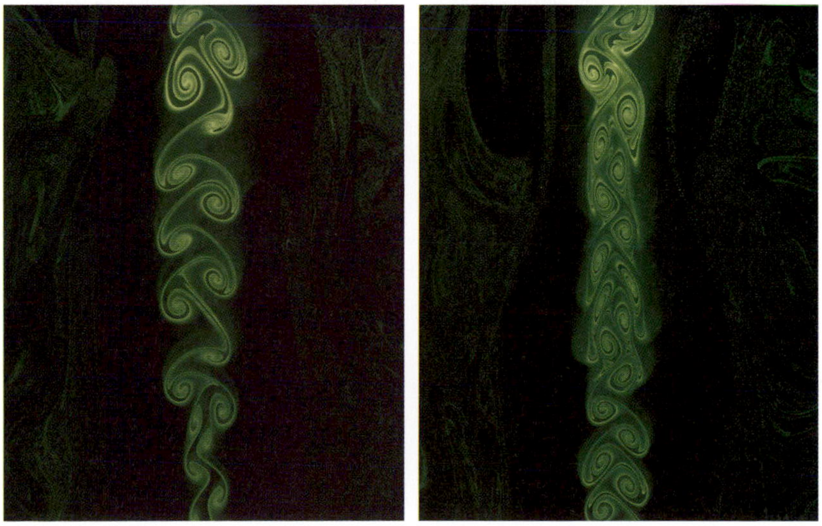

20

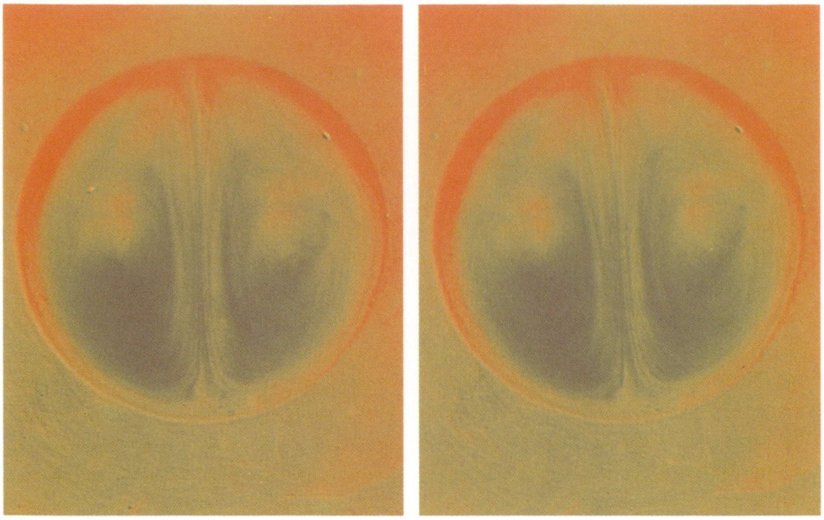

21

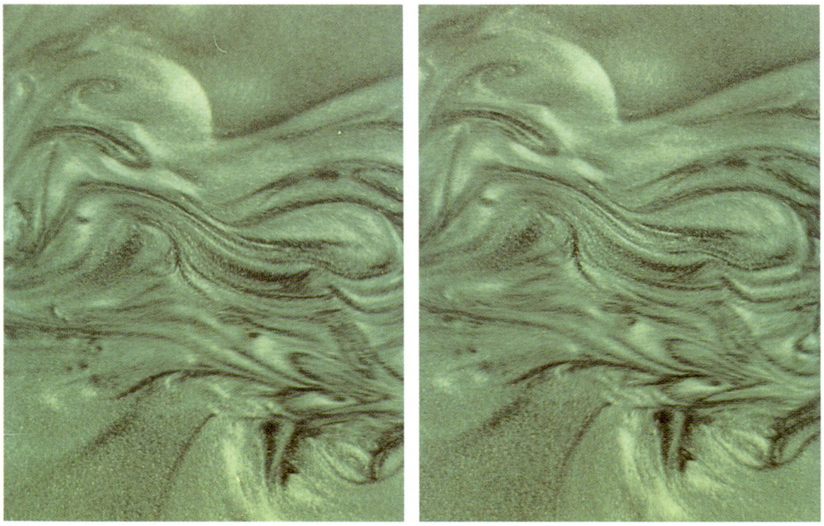

22

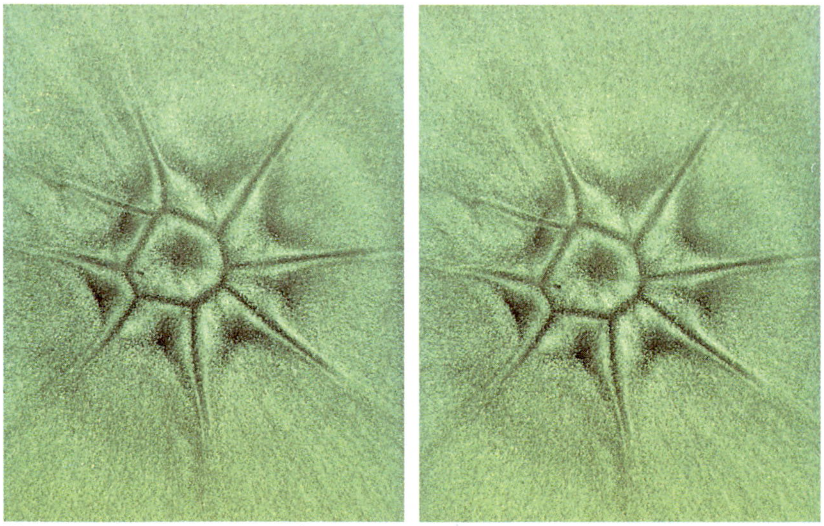

23

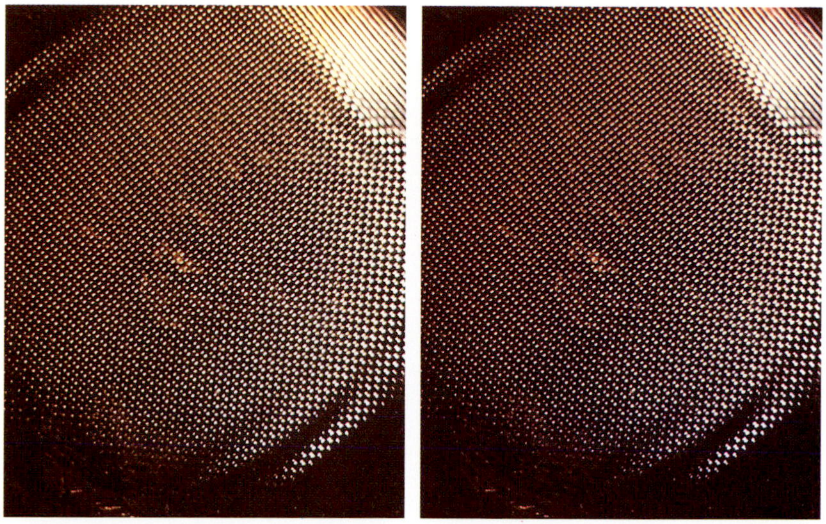

24

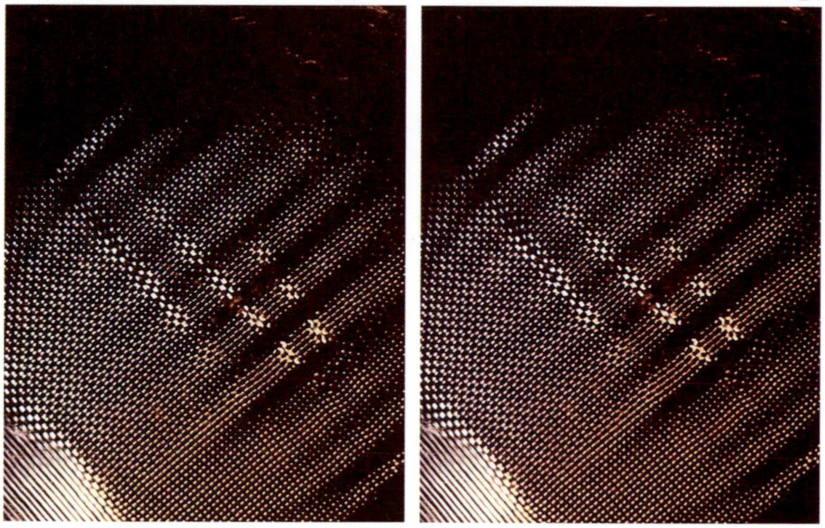

25

26

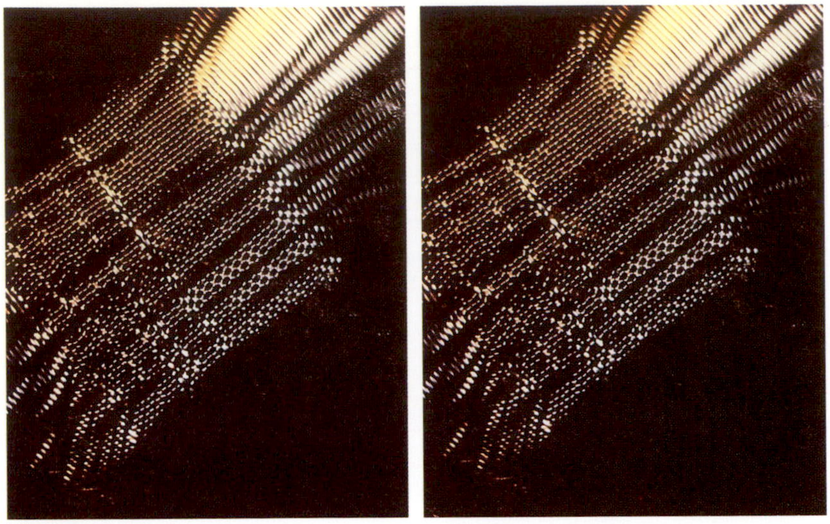

27

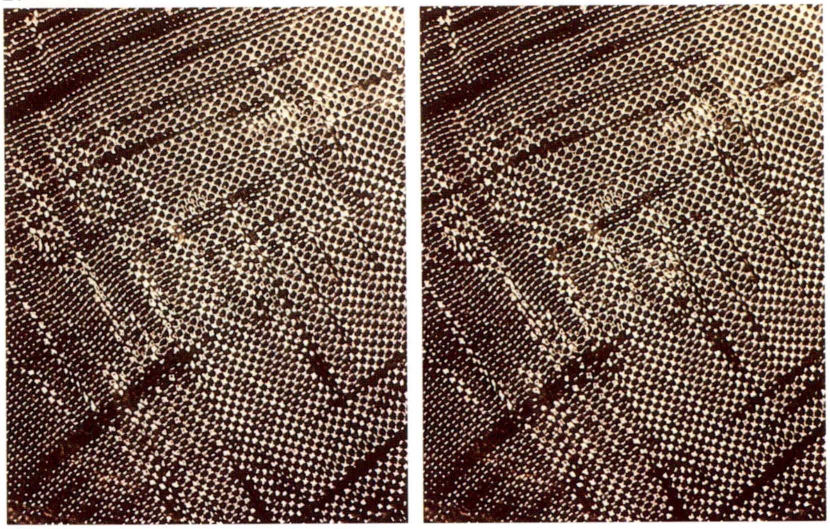

28

29

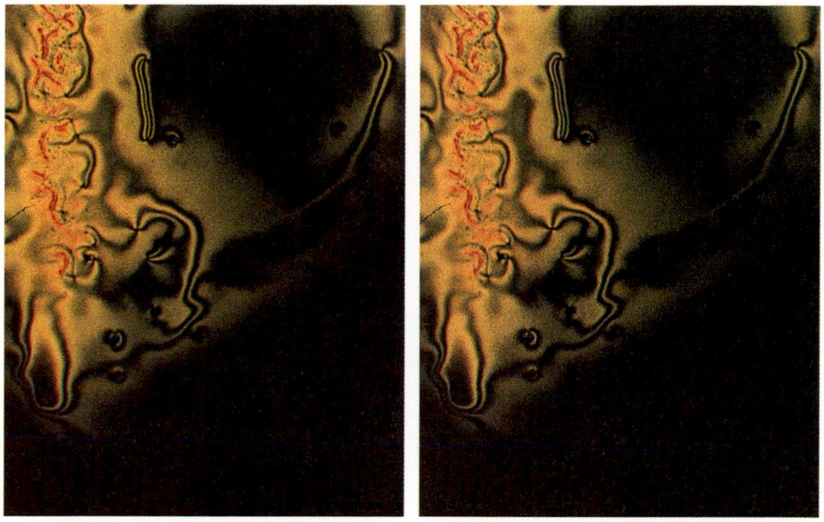

30

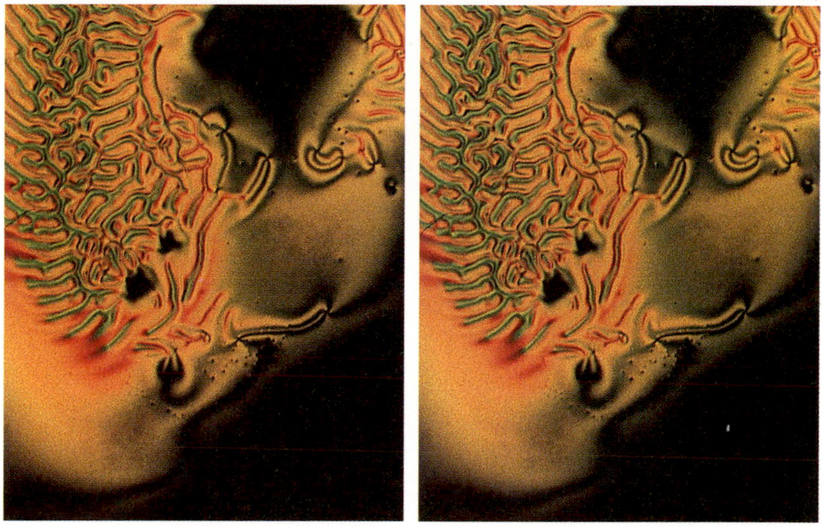

31

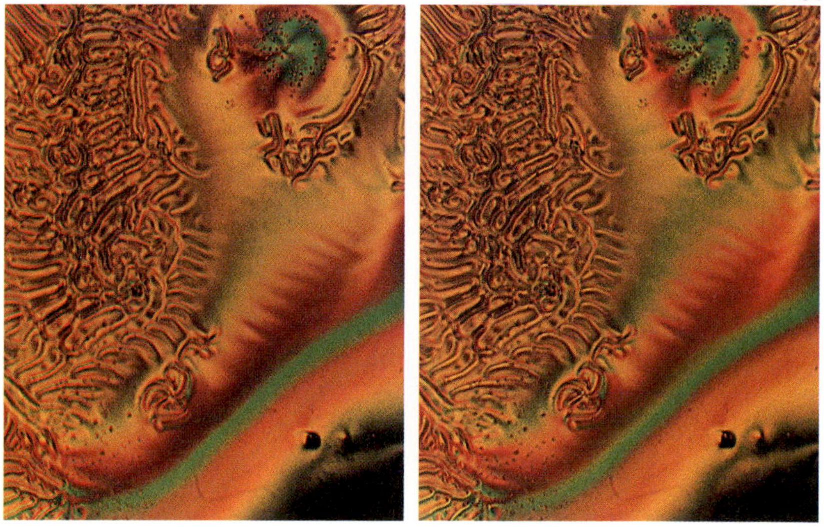

32

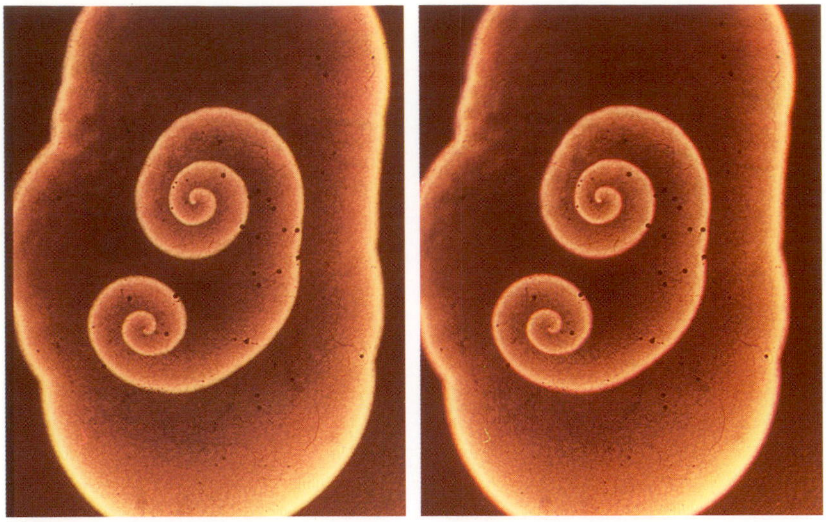

33

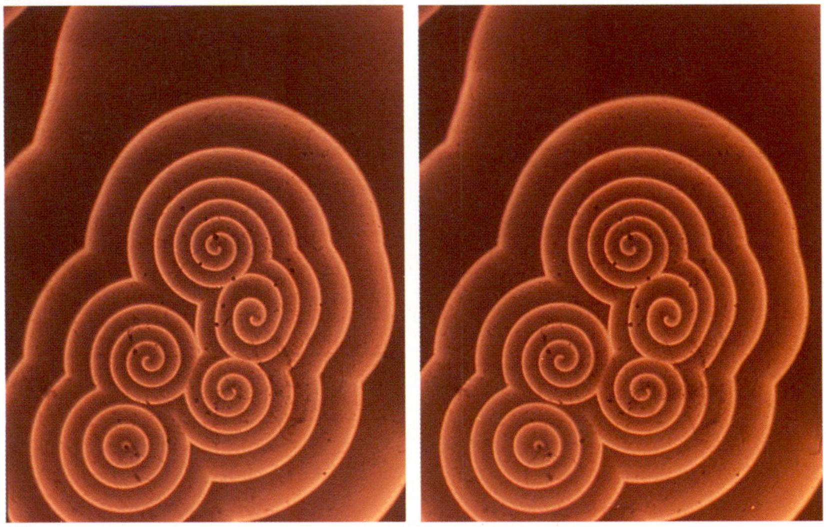

34

35

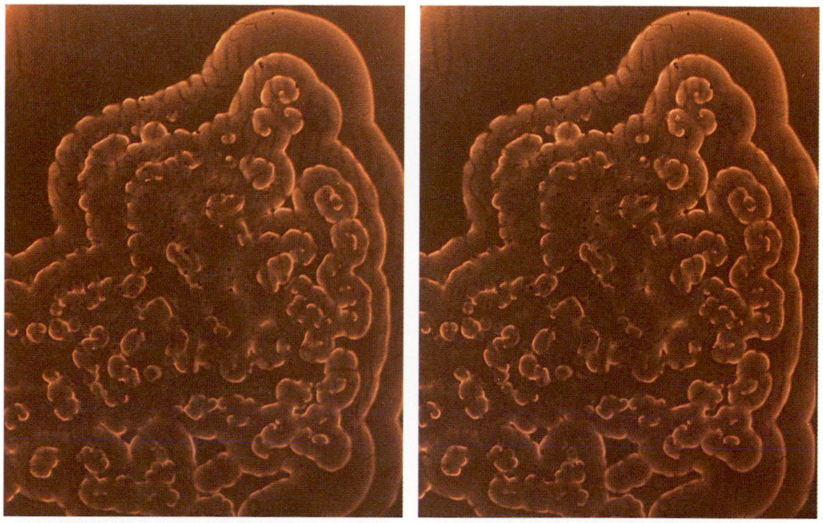

36

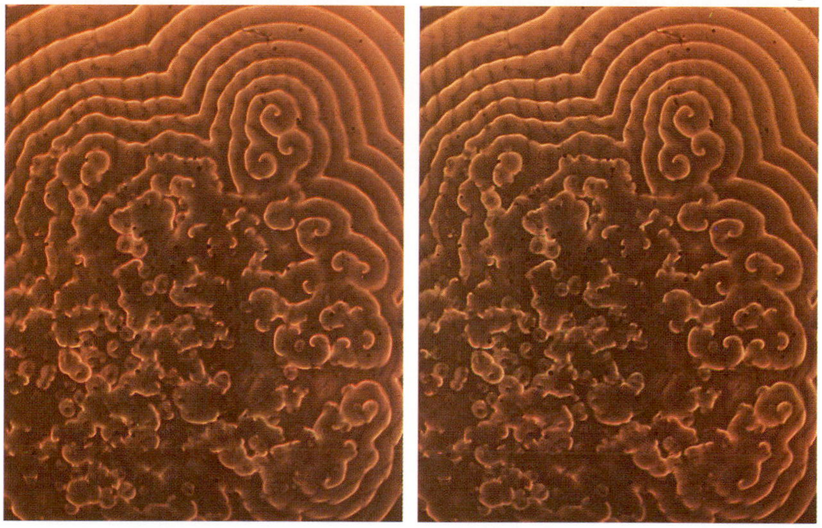

37

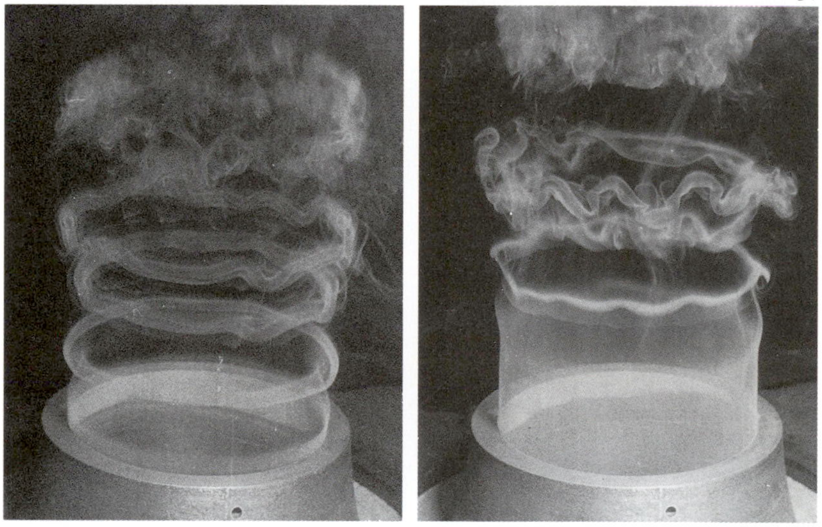

38

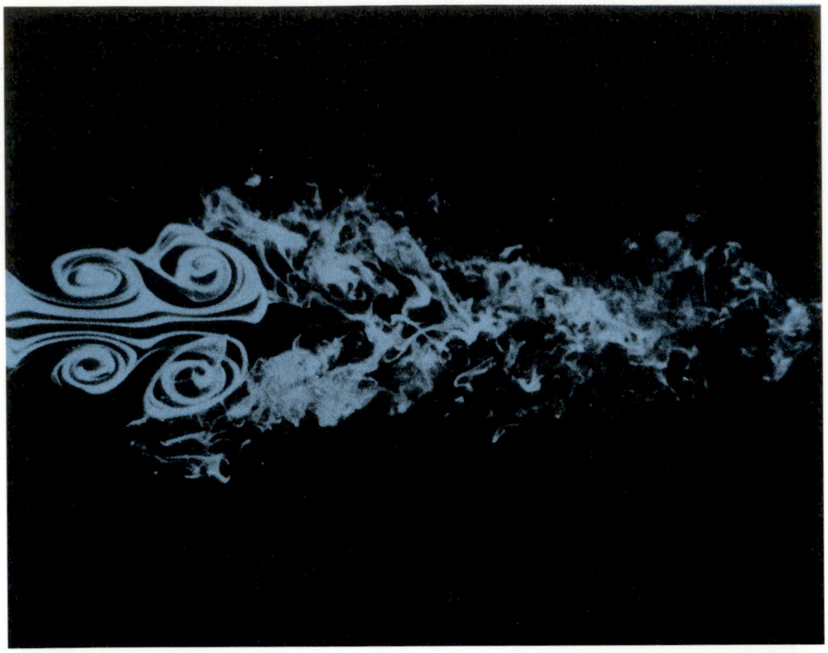
39

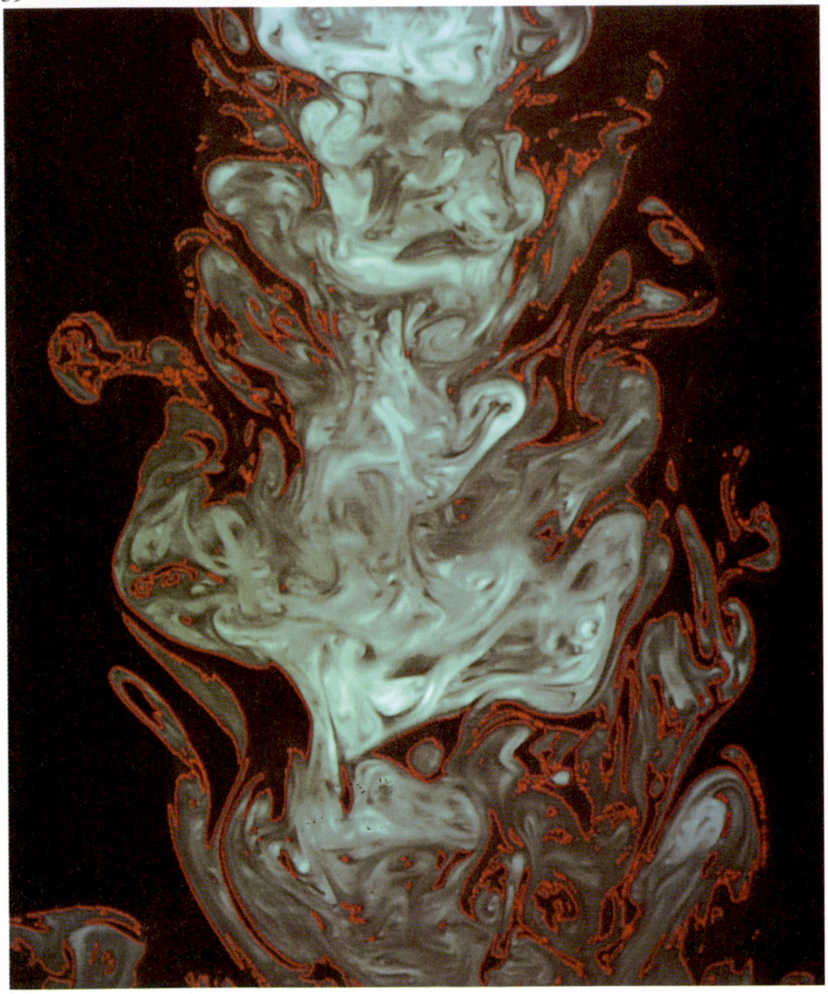
40

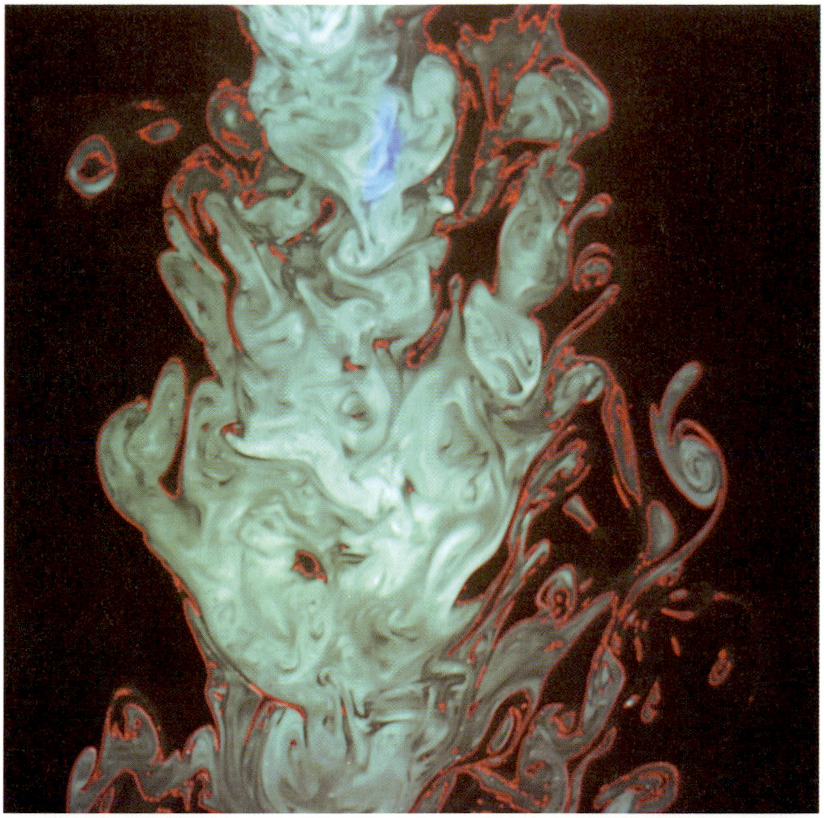

41

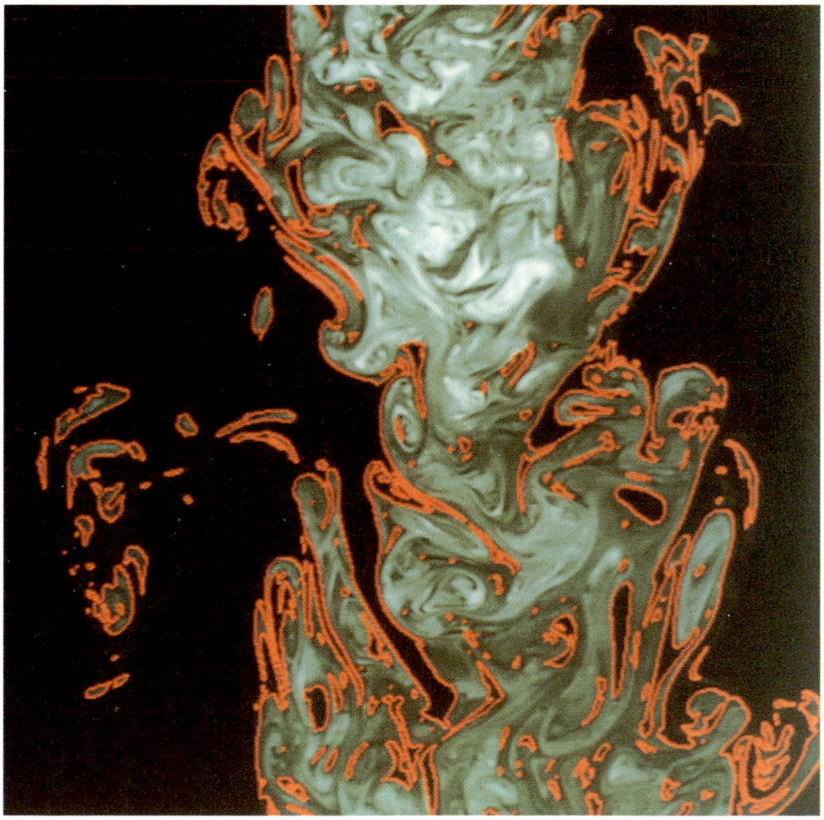

42

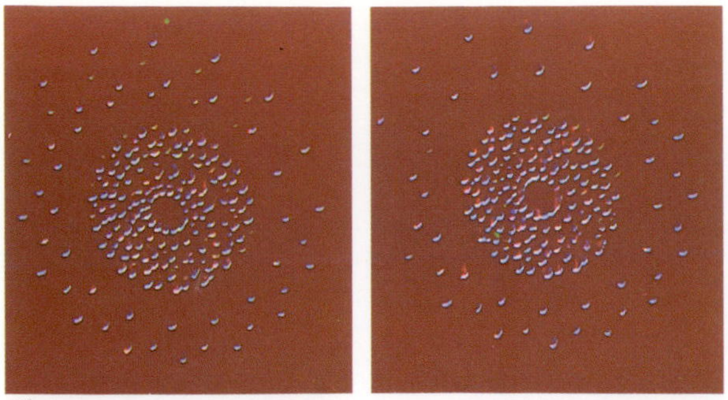

43

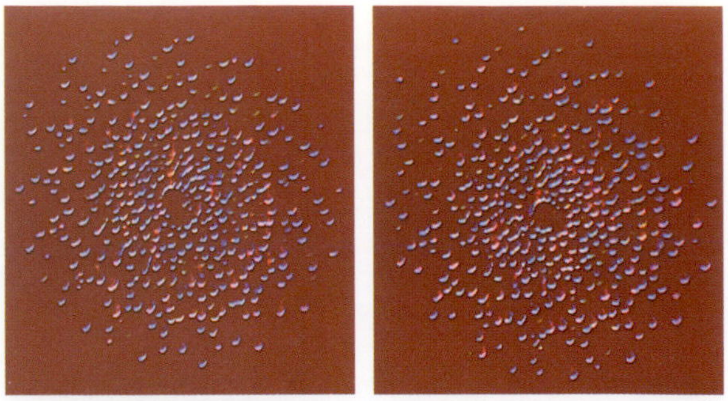

44

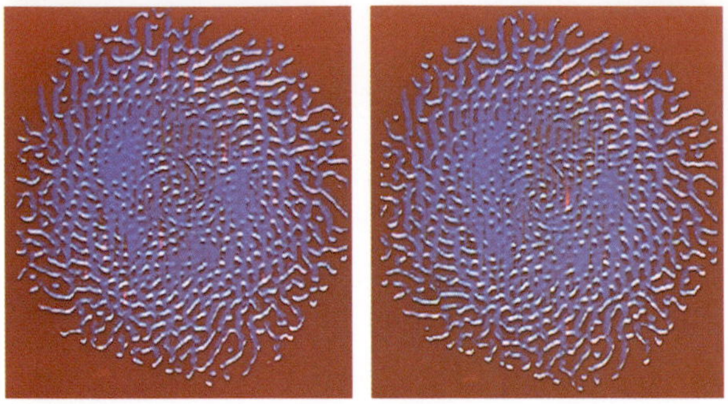

45

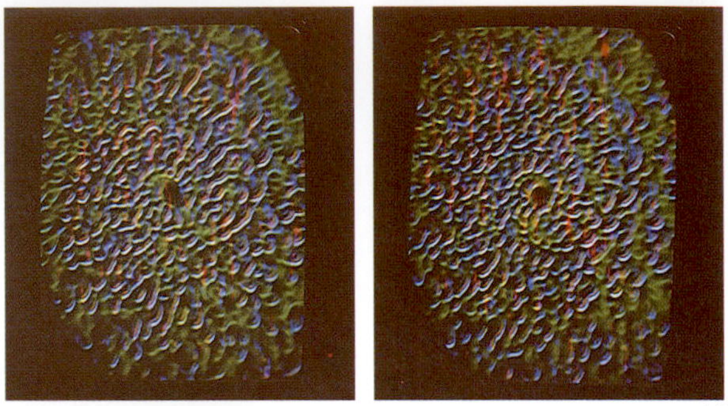

46

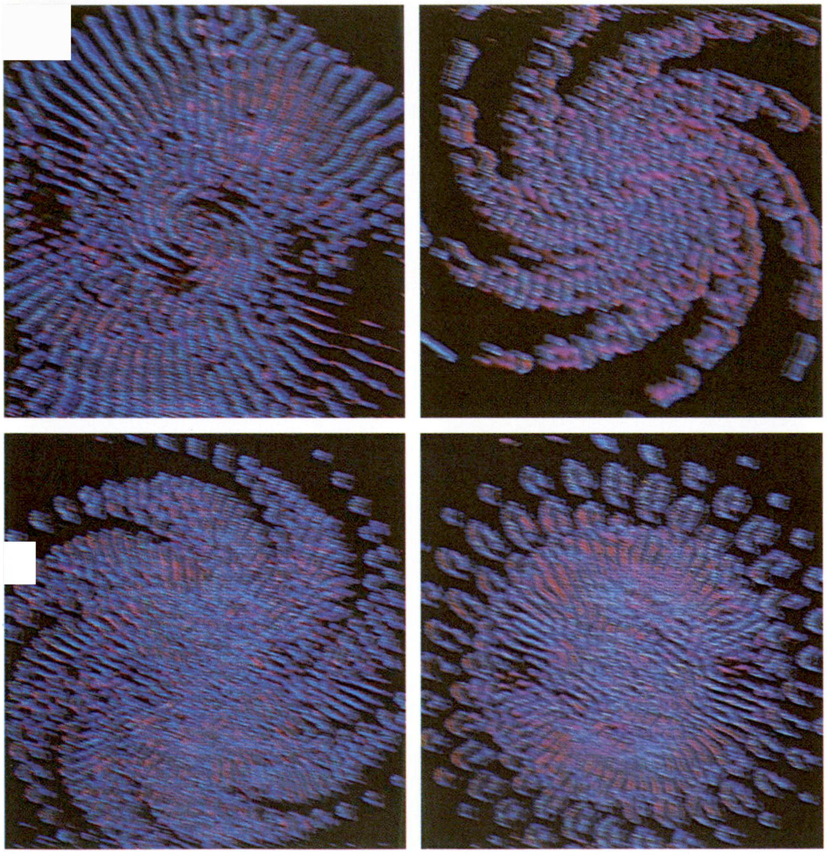

47

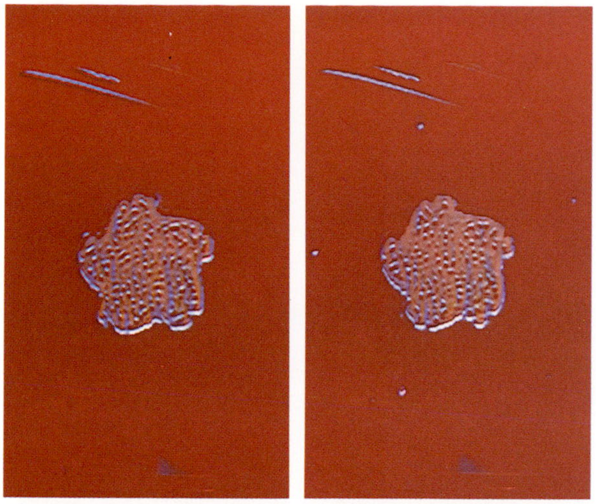

48

49

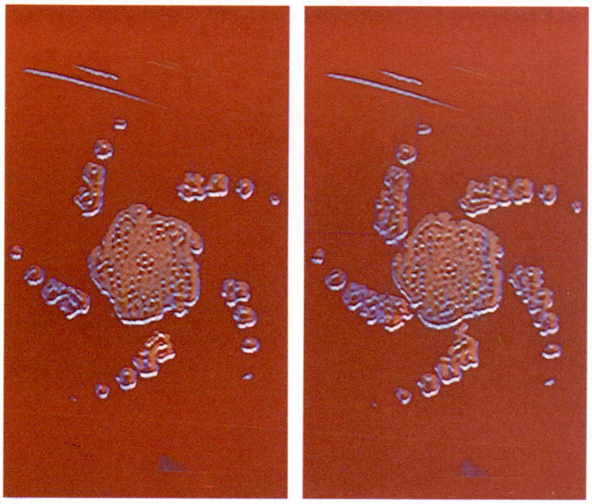

50

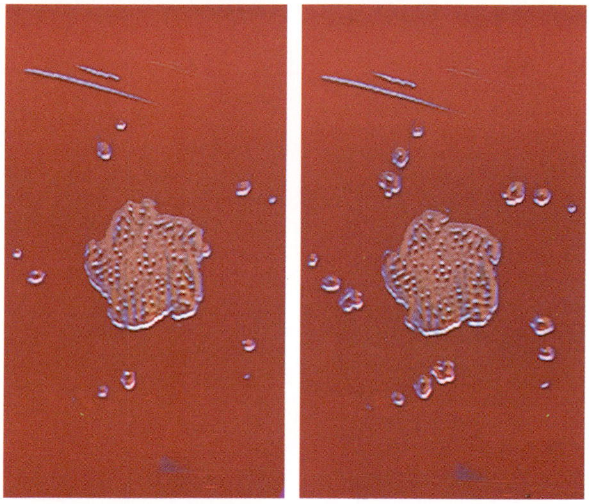

51

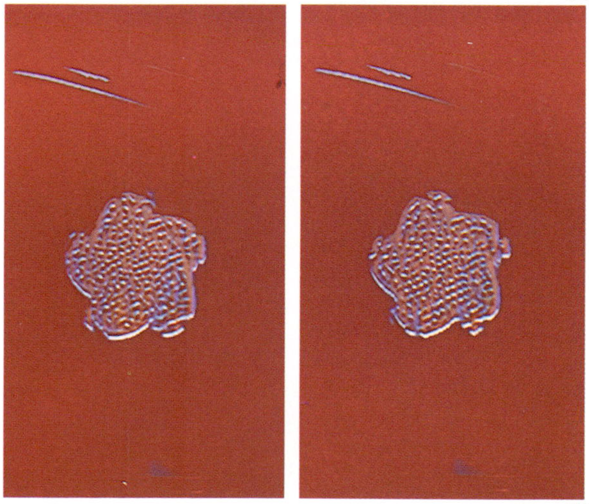

52

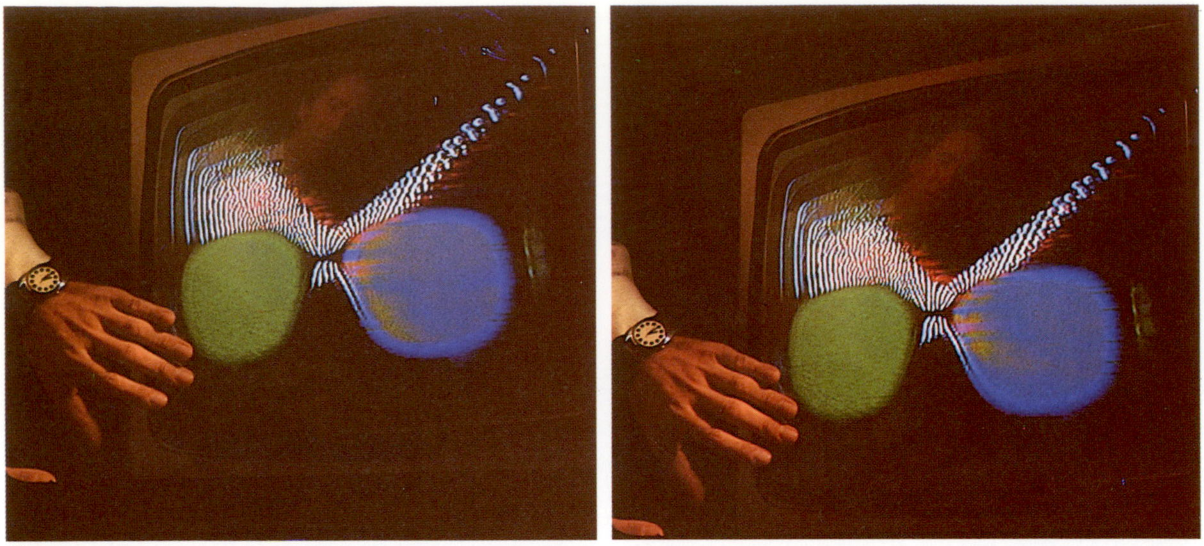

53

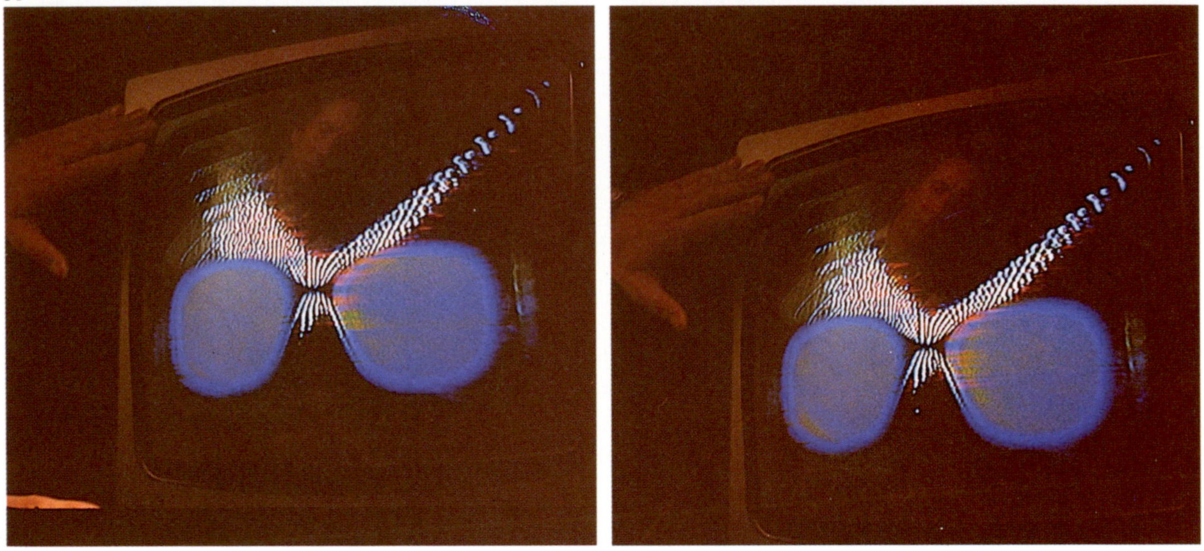

54

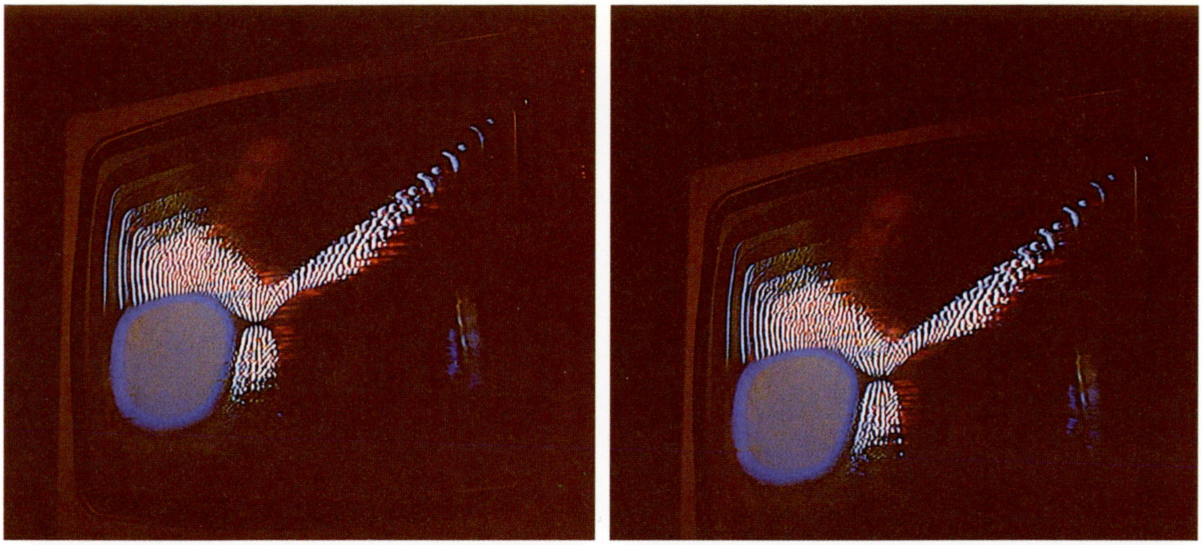

55

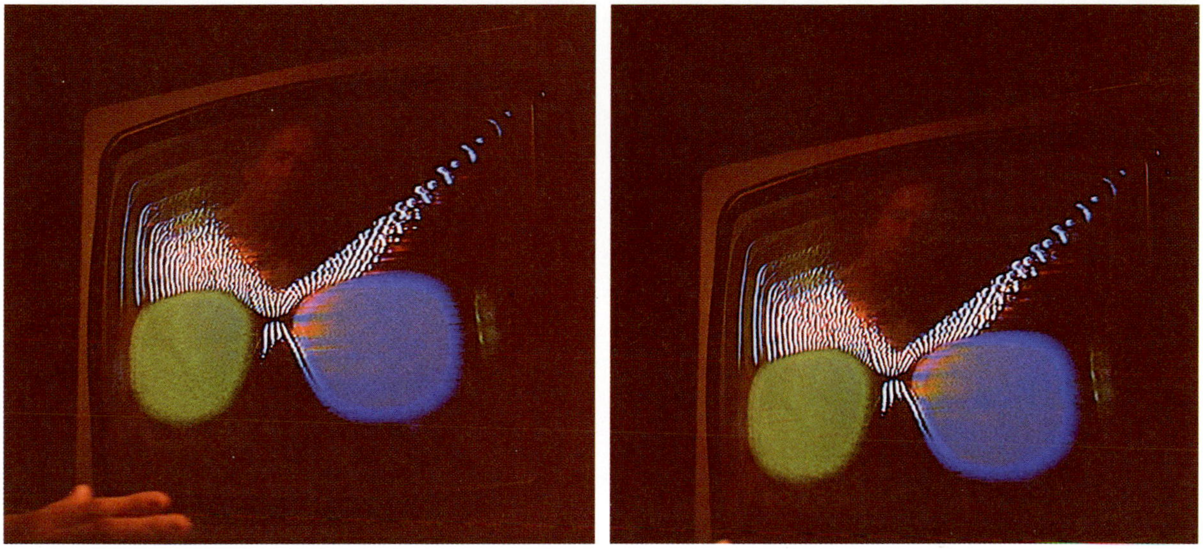

56

57

58

59

60

61

62

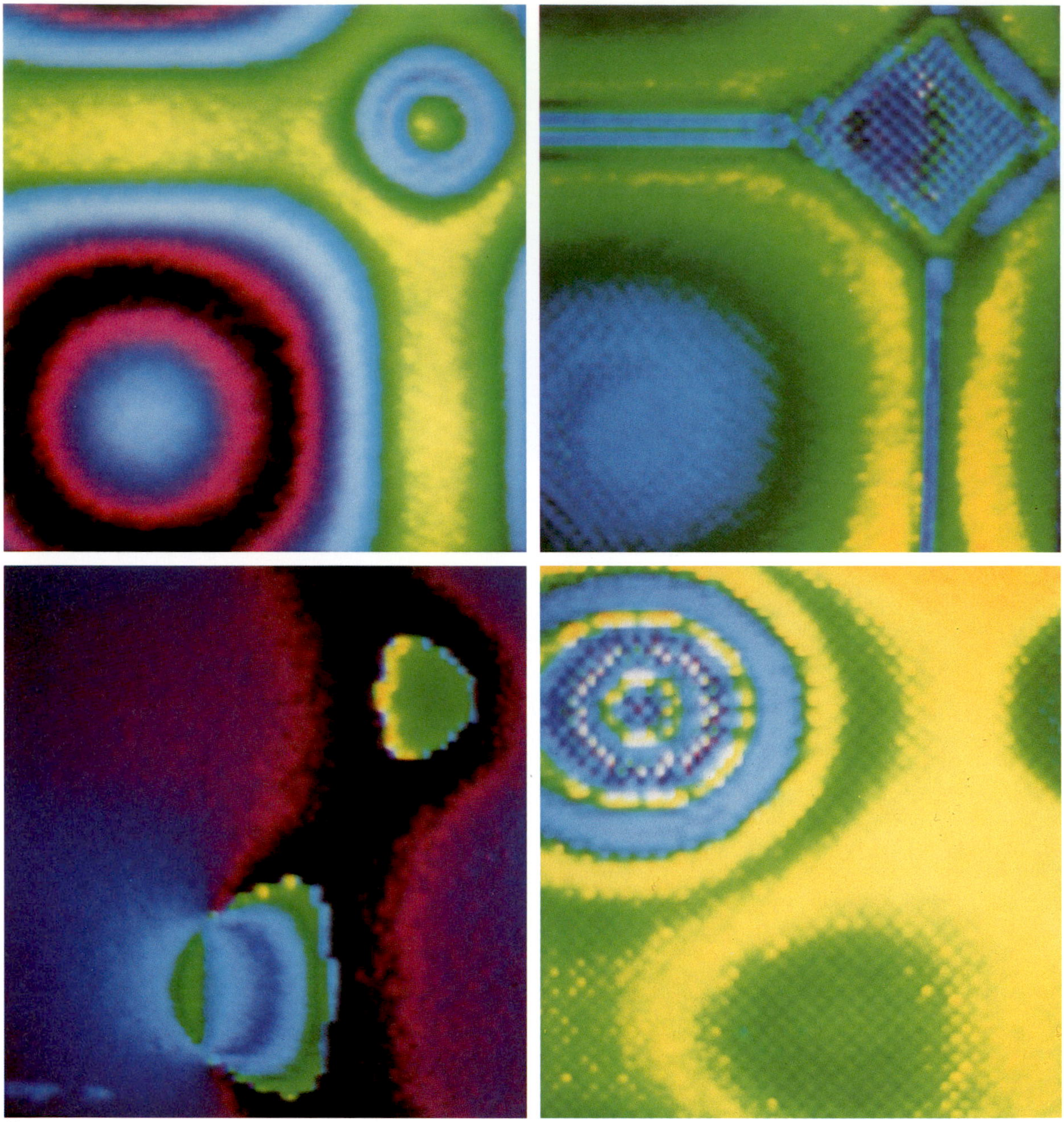

63

Literature

Chapter 1

1.1 L.I. Mandelstam: *Lectures on the Theory of Oscillations* (Nauka, Moscow 1972) [in Russian]

1.2 B. van der Pol: Radio Rev. **1**, 701 (1920)

1.3 A.A. Andronov, A.A. Vitt, C.E. Khaikin: *The Theory of Oscillations* (Fizmatgiz, Moscow 1959) [in Russian]

1.4 A.V. Gaponov-Grekhov, M.I. Rabinovich: Usp. Fiz. Nauk **128**, 579 (1979)

1.5 J.W.S. Rayleigh: *Theory of Sound* (MacMillan, London 1926)

1.6 E.M. Lifshitz, L.P. Pitaevski: *Physical Kinetics* (Pergamon, Oxford 1981)

1.7 R. Courant, K.O. Friedrichs: *Supersonic Flow and Shock Waves* (Interscience, New York 1948)

1.8 J.B. Benjamin, M.J. Lighthill: Proc. R. Soc. Lond. **A224**, 448 (1954)

1.9 M.A. Aizermann: *Lectures on the Theory of Automatic Regulation* (Fizmatgiz, Moscow 1958) [in Russian]

1.10 S.A. Akhmanov, R.V. Khokhlov: *Problems of Nonlinear Optics* (VINITI, Moscow 1964) [in Russian], see also
V.S. Butylkin, A.A. Kaplan, Y.G. Khronopulo, E.I. Yakubovich: *Resonant Nonlinear Interaction of Light with Matter* (Springer, Berlin, Heidelberg, New York 1989)

1.11 L. Blombergen: *Nonlinear Optics* (Harvard Univ. Benjamin, New York 1965)

1.12 V.M. Fain, Ya.I. Khanin: *Quantum Radiophysics* (Sov. Radio, Moscow 1965) [in Russian]

1.13 C.H. Tawns, A.L. Schawlow: Phys. Rev. **112**, 1940 (1958)

1.14 G.A. Askar'yan: Zh. Eksp. Teor. Fiz. **42**, 1567 (1962)

1.15 B.Ya. Zel'dovich, O.Yu. Nosach, V.I. Papovichev, V.V. Rogal'skii, Ph. S. Phaizullov: Vestn. Mosk. Fiz. Astr. **4**, 137 (1978)

1.16 V.N. Oraevsky, R.Z. Sagdeev: Zh. Tekhn. Fiz. **32**, 1291 (1962)

1.17 R.Z. Sagdeev, A.A. Galeev: in *Nonlinear Plasma Theory*, J. O'Neil, D. Book (Eds.) (Bemjamin, New York 1969)

1.18 A.A. Vedenov, Ye.P. Velikhov, R.Z. Sagdeev: Yad. Sintez. Prilozhen. **3**, 1049 (1962) [in Russian]

1.19 V.E. Zakharov: Prikl. Mekh. Tekh. Fiz. **4**, 35 (1965)

1.20 B.B. Kadomtsev: in *Problems of Plasma Theory*, issue 4 (Gosatomizdat, Moscow 1964) pp. 188 [in Russian]

1.21 N.N. Bogolyubov, Yu.A. Mitropol'sky: *Asymptotic Methods in the Theory of Nonlinear Oscillations* (Fizmatgiz, Moscow 1963)

1.22 Sh. Ma: *Modern Theory of Critical Phenomena* (Benjamin, New York 1976)

1.23 Ya.B. Zel'dovich, Yu.P. Raizer: *Physics of Shock Waves and High-Temperature Hydrodynamic Phenomena* (Nauka, Moscow 1966), see also
Yu.P. Raizer: *Gas Discharge Physics* (Springer, Berlin, Heidelberg, New York 1991)

1.24 H. Leaute: J. l'Ecole Polytechn. **55**, 1 (1885)

Chapter 2

2.1 M.I. Rabinovich, D.I. Trubetskov: *Introduction to the Theory of Oscillations and Waves* (Kluwer, Amsterdam 1989)
2.2 G.F. Gauze, A.A. Vitt: Izv. AN SSSR **7**, 1551 (1936)
2.3 E. Fermi: Z. Phys. **71**, 250 (1931)
2.4 A.A. Vitt, G.S. Gorelik: Zh. Tekhn. Fiz. **3**, 294 (1933)
2.5 R.V. Khokhlov: Raditotekhn. Electron. **6**, 1116 (1961)
2.6 P. Paradoksov: Usp. Fiz. Nauk **89**, 707 (1966) [English: Sov. Phys. Usp. **9**, 618 (1967)]
2.7 A.A. Andronov: *Collected Works* (AN SSSR, Moscow 1955) p. 32 [in Russian]
2.8 P.A. Franken, A.E. Hill, C.W. Peters, J. Weinreich: Phys. Rev. Lett. **7**, 118 (1961)
2.9 M. Henon, C. Heiles: Astron. J. **69**, 73 (1964)
2.10 B.B. Kadomtsev: *Collective Phenomena in Plasma* (Nauka, Moscow 1976)
2.11 E. Fermi, J. Pasta, S. Ulam: in *E. Fermi - Scientific Works*, Vol. 2 (Nauka, Moscow 1972) p. 643
2.12 V.E. Zakharov: Prikl. Mekh. Tekhn. Fiz. **4**, 35 (1965)
2.13 K.A. Gorshkov, L.A. Ostrovsky, V.V. Papko, A.S. Pikovsky: Phys. Lett. **74**, 223 (1979)
2.14 Yu.A. Danilov: in *Nonlinear Waves 1 - Dynamics and Evolution*, A.V. Gaponov-Grekhov, M.I. Rabinovich, J. Engelbrecht (Eds.) (Springer, Berlin, Heidelberg, New York 1989) p. 2
2.15 A. Poincaré: *Calcul des probabilités* (Gautier-Villard, Paris 1912)
2.16 S. Smale: Matematika **11**, 88 (1967)
2.17 L.P. Shil'nikov: Mat. Sborn. **74**, 378 (1967)
2.18 Yu.I. Neimark: *The Method of Point Maps in Nonlinear Oscillation Theory* (Nauka, Moscow 1972) [in Russian]
2.19 M.D. Kruskal, N.J. Zabusky: Phys. Rev. Lett. **15**, 240 (1965); I.A. Kunin: *Theory of Elastic Media with Microstructure* (Springer, Berlin, Heidelberg, New York 1982) Chap. 5
2.20 V.E. Zakharov: Zh. Eksp. Teor. Fiz. **65**, 219 (1973) [English: Sov. Phys. JETP **38**, 108 (1974)
2.21 R.K. Dodd, J.C. Eilbeck, J.D. Gibbon, H.C. Morris: *Solitons and Nonlinear Wave Equations* (Academic, New York 1984)
2.22 D.W. McLaughlin, A.C. Scott: in *Solitons in Action*, K. Longren, A. Scott (Eds.) (Academic, New York 1978) p. 201
2.23 K.A. Gorshkov, L.A. Ostrovsky, V.V. Papko: Zh. Eksp. Teor. Fiz. **71**, 585 (1976) [English: Sov. Phys. JETP **44**, 306 (1976)
2.24 E.N. Pelinovsky: in *Nonlinear Waves. Propagation and Interaction* (Nauka, Moscow 1981) p. 187 [in Russian]
2.25 H.C. Yuen, B.N. Lake: *Solitons in Action* K. Longren, A. Scott (Eds.) (Academic, New York 1978)
2.26 R.D. Ray: J. Acoust. Soc. Amer. **3**, 222 (1931)
2.27 R.Z. Sagdeev: in *Problems in Plasma Theory*, Issue 4 (Atomizdat, Moscow 1964) p. 20 [in Russian]
2.28 S.I. Syrovatsky: Usp. Fiz. Nauk **62**, 247 (1957)
2.29 L.D. Landau, E.M. Lifshitz: *Electrodynamics of Continuous Media* (Pergamon, Oxford 1960)
2.30 A.V. Gaponov-Grekhov, G.I. Freidman: Zh. Eksp. Teor. Fiz. **39**, 957 (1959)
2.31 A.L. Fabrikant, V.V. Zheleznyakov: Zh. Eksp. Teor. Fiz. **82**, 1366 (1982)
2.32 R.V. Polovin: Diff. Uravn. **1**, 499 (1965)
2.33 P.D. Lax: Com. Pure Appl. Math., **21**, 467 (1969)
2.34 S.V. Manakov, S.P. Novikov, L.P. Pitaevsky, V.E. Zakharov: *Theory of Solitons* (Nauka, Moscow 1980) [in Russian]
2.35 T. Maxworthy, L.G. Redekopp: Icarus **29**, 261 (1976)
2.36 V.I. Petviashvili: Pis'ma Zh. Eksp. Teor. Fiz. **32**, 632 (1980)
2.37 M.V. Nezlin: Usp. Fiz. Nauk **150**, 3 (1986); M.V. Nezlin, E.B. Snezhkin: *Rossby*

Vortices, Spiral Waves and Solitons (Springer, Berlin, Heidelberg, New York 1992)
2.38 V.E. Zakharov, E.A. Kuznetsov: Zh. Eksp. Teor. Fiz. **66**, 594 (1974)
2.39 B.B. Kadomtsev, V.I. Petviashvili: Dokl. AN SSSR **192**, 753 (1970)
2.40 V.I. Petviashvili: Fiz. Plasmy **2** , 460 (1976)
2.41 E.A. Kuznetsov, S.K. Turitsyn: Zh. Eksp. Teor. Fiz. **36**, 190 (1982)
2.42 J.E. Brown, A.D. Jackson: *The Nucleon-Nucleon Interaction* (North-Holland, New York 1976)
2.43 R.D. Parmentier: in *Solitons in Action*, K. Longren, A. Scott (Eds.) (Academic, New York 1978) p. 173
2.44 L.A. Ostrovsky: Zh. Eksp. Teor. Fiz. **51**, 1189 (1966)
2.45 H.G. Yuen, W.E. Ferguson: Phys. Fluids **21**, 1275 (1978)
2.46 A.H. Luther: in *Solitons*, R.K. Bullough, P.J. Caudrey (Eds.) (Springer, Berlin, Heidelberg, New York 1980) p. 355
2.47 A.S. Davydov: in *Nonlinear Waves. Propagation and Interaction* (Nauka, Moscow 1981) p. 42 [in Russian]
2.48 S.F. Krylov, V.V. Yan'kov: Zh. Eksp. Teor. Fiz. **79**, 82 (1980)
2.49 L.I. Mandelstam, A.A. Papaleksi: Zh. Eksp. Teor. Fiz. **4**, 117 (1934)
2.50 M.I. Rabinovich, A.A. Rozenblum: Dokl. AN SSSR **213**, 1276 1973) [English: Sov. Phys. Dokl. AN SSSR **18**, 803 1974)
2.51 S.V. Kiyashko, M.I. Rabinovich, V.P. Reutov: Zh. Tekhn. Fiz. **42**, 2458 (1973)
2.52 M.I. Rabinovich: Izv. VUZov, Radiofizika **17**, 475 (1974)
2.53 M.Z. Gak: Izv. AN SSSR – Fiz. Atmosfery i Okeana **17**, 201 (1981) [in Russian]
2.54 J.E. Marsden, M. McCracken: *The Hopf Bifurcation and its Applications* (Springer, Berlin, Heidelberg, New York 1976)
2.55 E.C. Zeeman: *Catastrophe Theory - Selected Papers* (1972–1977) (Wesley, Massachusetts 1977)
2.56 M.A. Leontovich: Sov. Phys. Usp. **126**, 673 (1978)
2.57 C.V. Raman, K.S. Krishnan: Phil. Mag. **5**, 498 (1928)
2.58 S.A: Akhmanov, R.V. Khoklov: *Problems of Nonlinear Optics*, (VINITI, Moscow 1964)
2.59 N. Blombergen: *Nonlinear Optics* (Benjamin, New York 1965)
2.60 V.N. Tsytovich: *Nonlinear Effects in Plasmas* (Nauka, Moscow 1970) [in Russian]
2.61 V.A. Baily, P.F. Martin: Phil. Mag. **18**, 369 (1934)
2.62 Y.V. Gulyaev, P.E. Zil'berman: Fiz. Tekh. Poluprovodn. **5**, 126 (1971) [English: Sov. Phys. Semicond. **5**, 103 (1971)]
2.63 E.I. Yakubovich: Zh. Eksp. Teor. Fiz. **56**, 676 (1969); [English: Sov. Phys. JETP **29**, 370 (1969)]
2.64 I.L. Bershtein, V.A. Rogachev: Izv. Vyssh. Uchebn. Zaved. Radiofiz. **13**, 33 (1970)
2.65 T.W. Hänsch: Physics Today **30**, 34 (1977)
2.66 B.Ya. Zel'dovich, V.I. Popovichev, V.V. Rogal'skii, Ph.S. Fizullov:Pis'ma Zh. Eksp. Teor. Fiz. **15**, 160 (1972); [English: Sov. Phys. JETP Lett. **15**, 109 (1972)]
2.67 V.I. Bespalov, G.A. Pasmanik: *Nonlinear Optics and Adaptive Laser Systems* (Nauka, Moscow 1986) p. 135 [in Russian]
2.68 B.Ya Zel'dovich: Vestnik Mosk. Univ. Fiz. Astron. **4**, 137 (1978)
2.69 M.J. Lighthill: J. Inst. Math. Appl. **1**, 269 (1965)
2.70 A.A. Galeev, R.Z. Sagdeev, Yu.S. Sigov, V.D. Shapiro, B.A. Shevchenko: Fiz. Plazmy **1**, 10 (1975); [English: Sov. J. Plasma Phys. **1**, 5 (1975)]
2.71 A.G. Litvak, V.I. Talanov: Fiz. Plazmy **10**, 539 (1967)
2.72 T.B. Benjamin, F.E. Feir: J. Fluid Mech. **27**, 417 (1967)
2.73 V.I. Talanov: Pis'ma Zh. Eksp. Teor. Fiz. **2**, 218 (1965); [English: Sov. Phys. JETP Lett. **2**, 138 (1965)]
2.74 L.A. Ostrovskii: Zh. Eksp. Teor. Fiz. **51**, 1189 (1966); [English: Sov. Phys. JETP **24**, 797 (1967)]
2.75 H.C. Yen, B.M. Lake: Phys. Fluids **18**, 956 (1975); Advances Appl. Mech. **22**, 67 (1982)

2.76 V.E. Zakharov, A.M. Rubenchik: Zh. Eksp. Teor. Fiz. **65**, 997 (1973); [English: Sov. Phys. JETP **38**, 494 (1974)]
2.77 M.I. Rabinovich, A.L. Fabrikant: Zh. Eksp. Teor. Fiz. **77**, 617 (1979); [English: Sov. Phys. JETP **50**, 311 (1979)]

Chapter 3

3.1 B. van der Pol, J. van der Mark: Natura **120**, 363 (1927)
3.2 T. Rikitake: Proc. Cambr. Phil. Soc. **54**, 89 (1958)
3.3 A.S. Alekseev: *Sbornik pam'yati A.A. Andronova* (AN SSSR, Moscow 1955)
3.4 E. Lorenz: J. Atmos. Sci. **20**, 130 (1963)
3.5 H. Haken: Phys. Lett. **53A**, 77 (1975); R. Graham: Phys. Lett. **58A**, 440 (1976)
3.6 M.I. Rabinovich: Sov. Phys. Usp. **21**, 443 (1978)
3.7 D.D. Rutov, G.V. Stupakov: Pis'ma Zh. Eksp. Teor. Fiz. **26**, 186 (1977)
3.8 V.D. Il'in, A.N. Il'ina: Zh. Eksp. Teor. Fiz. **72**, 983 (1977)
3.9 E.F. Jaeger, A.J. Lichtenberg, M.A. Lieberman: Plasma Phys. **14**, 1073 (1972)
3.10 G.M. Zaslavski, B.V. Chirikov: Sov. Phys. Usp. **105**, 3 (1971)
3.11 S.V. Kiyashko, A.S. Pikovsky, M.I. Rabinovich: Radiotekh. Elektr. **25**, 336 (1980)
3.12 J. Wisdom, S.J. Peale, F. Mignard: Icarus **58**, 137 (1984)
3.13 A.J. Lichtenberg, M.A. Lieberman: *Regular and Stochastic Motion*, (Springer, Berlin, Heidelberg, New York 1983); W. Dittrich and M. Reuter: *Advanced Classical and Quantum Dynamics - From Classical Paths to Path Integrals*, (Springer, Berlin, Heidelberg, New York 1992)
3.14 V.I. Arnol'd: Dokl. AN SSSR **156**, 9 (1964) [in Russian]
3.15 B.V. Chirikov: Phys. Rep. **52**, 265 (1979)
3.16 N.S. Krylov: *Works on the Justification of Statistical Physics* (AN SSSR, Moscow 1950) [English: (Princeton University Press, Princeton 1979)]
3.17 M. Born: Z. Phys. **153**, 372 (1958)
3.18 Ya.G. Sinai: Priroda **3**, 72 (1981)
3.19 D.V. Anosov, V.I. Arnol'd, S.P. Novikov, Ya.G. Sinai (Eds.) *Encyclopedia of Mathematical Science - Dynamical Systems* (Springer, Berlin, Heidelberg, New York 1987)
3.20 H.F. Creveling, J.F. de Paz, J.Y. Baladi, R.J. Schoenhals: J. Fluid Mech. **67**, 65 (1965)
3.21 V.S. Afraimovich, V.V. Bykov, L.P. Shil'nikov: Trans. Moscow Math. Soc. **44**, 153 (1983)
3.22 M.L. Cartwright, G.E. Littlewood: J. London Math. Soc. **20**, 180 (1945)
3.23 C. Pezeshki, E.H. Dowell: Physica **D32**, 194 (1988)
3.24 M.P. Kennedy, L.O. Chua: IEEE Trans. Circ. Syst. **CAS-33**, 974 (1986)
3.25 C. Grebogi, E. Ott, J.A. Yorke: Physica **24D**, 243 (1987)
3.26 A.S. Pikovsky: Radiophys. Quant. Electr. **29**, 389 (1986)
3.27 J.P. Eckmann: Rev. Mod. Phys. **53**, 643 (1981)
3.28 M.J. Feigenbaum: J. Stat. Phys. **19**, 25 (1978)
3.29 P. Berge, M. Debois, P. Manneville: J. Phys. Lett. **41**, 341 (1980)
3.30 P. Berge, Y. Pomeau, C. Vidal: *Order Within Chaos* (Wiley, New York and Herman, Paris 1984)
3.31 S. Newchause, D. Ruelle, F. Takens: Commun. Math. Phys. **64**, 35 (1978)
3.32 M.J. Feigenbaum, L. Kadanoff, S. Shenker: Physica **D5**, 370 (1982)
3.33 R. Adler, G. Konheim, M.H. McAndrew: Am. Math. Soc. **114**, 309 (1965)
3.34 R. Bowen: Trans. Am. Math. Soc. **153**, 401 (1971)
3.35 F. Takens: Lect. Notes Math., Vol. 898 (Springer, Berlin, Heidelberg, New York 1980) p. 336
3.36 F. Takens: in *Nonlinear Dynamics and Turbulence*, G.I. Barenblatt, G. Iooss, D.D. Joseph (Eds.) (Pitman, Boston 1983) p.314
3.37 V.I. Oseledec: Trudy Mosk. Mat. Obshch. **19**, 179 (1968) [in Russian]
3.38 F. Ledrappier: Comm. Math. Phys. **81**, 229 (1981)

3.39 J.P. Eckmann, D. Ruelle: Rev. Mod. Phys. **57**, 617 (1985)
3.40 P. Grassberger, I. Procaccia: Physica **D9**, 189 (1983)
3.41 A. Ben-Mizrachi, I.Procaccia: Phys. Rev. **A29**, 975 (1984)
3.42 V.S. Afraimovich, A.M. Reiman: in *Nonlinear Waves 2*, A.V. Gaponov-Grekhov, M.I. Rabinovich, J. Engelbrecht (Eds.) (Springer, Berlin, Heidelberg, New York 1989)
3.43 G. Mayer-Kress (Ed.) *Dimension and Entropies in Chaotic Systems* (Springer, Berlin, Heidelberg, New York 1987)
3.44 T.S. Parker and L.O. Chua: *Practical Numerical Algorithms for Chaotic Systems* (Springer, Berlin, Heidelberg, New York 1989)

Chapter 4

4.1 N. Wiener, A. Rosenbluth: Arch. Inst. Cardiol. Mex. **16**, 205 (1946)
4.2 A.M. Turing: Phil. Trans. Roy. Soc. London **B237**, 37 (1952)
4.3 B.P. Belousov: *Abstracts of Papers on Radiation Medicine* (Medgiz, Moscow 1959) p. 145 [in Russian]
4.4 A.M. Zhabotinsky: *Concentration Self-Oscillations* (Nauka, Moscow 1974) [in Russian]
4.5 C. Normand, Y. Pomeau, M.G. Velarde: Rev. Mod. Phys. **49**, 591 (1977)
4.6 J.T. Stuart: J. Fluid Mech. **4**, 1 (1958)
4.7 P. Glansdorff, I. Prigogine: *Thermodynamics of Structures, Stability and Fluctuations* (Wiley, New York 1971)
4.8 H. Haken: *Synergetics* (Springer, Berlin, Heidelberg, New York 1978)
4.9 *Autowave Processes in Diffusive Systems* M.T. Grekhova (Ed.) (Inst. Prikl. Fiz. AN SSSR, Gorky 1981) [in Russian]
4.10 G. Nicolis, I. Prigogine: *Self-Organization in Non-Equilibrium Systems* (Wiley, New York 1977)
4.11 L.S. Polak, A.S. Mikhailov: *Self-Organization in Non-Equilibrium Physico-Chemical Systems* (Nauka, Moscow 1983) [in Russian]
4.12 K.J. Tomchik, P.N. Devreotie: Science **212**, 4493 (1981)
4.13 M.I. Rabinovich, D.I. Trubetskov: *Introduction to the Theory of Oscillations and Waves* (Reidel, Amsterdam 1989)
4.14 V.I. Petviashvili, O.Yu. Tsvelodub: Sov. Phys. Dokl. **23**, 117 (1978)
4.15 V.I. Petviashvili: in *Nonlinear Waves*, Gaponov-Grekhov (Ed.) (Nauka, Moscow 1979) p. 5
4.16 A. Yariv: *Quantum Electronics*, third edition (Wiley, New York 1989) p. 676
4.17 V.V. Barelko, I.I. Kurochka, A.G. Merzhanov, A.P. Schkadinskii: Chem. Eng. Sci. **33**, 805 (1978)
4.18 A.C. Scott: Rev. Mod. Phys. **47**, 320 (1975)
4.19 M.A. Allessie, F.I.N. Boake, F.I.G. Schopman: Circul. Res. **41**, 9 (1977)
4.20 V.I. Koroleva, G.D. Kuznetsova: in *Electric Activity of the Brain* (Nauka, Moscow 1971) p. 130 [in Russian]
4.21 A. Roshko: AIAA Journal **11**, 1349 (1976)
4.22 M.I. Rabinovich, M.M. Sushchik: in *Nonlinear Waves: Self-Organization* (Nauka, Moscow 1983) [in Russian]; Sov. Phys. Uspekhi **33**, 1 (1990)
4.23 C. Vidal, A. Pacault: in *Evolution of Order and Chaos*, H. Haken (Ed.) (Springer, Berlin, Heidelberg, New York 1982) p. 74
4.24 Y. Kuramoto, S. Koge: Prog. Theor. Phys. **66**, 1081 (1981)
4.25 V.I. Petviashvili, V.V. Yan'kov: in *Voprosy Teorii Plazmy*, Vyp. 14, B.B. Kadomtsev (Ed.) (Atomizdat, Moscow 1985) p.3 [in Russian]
4.26 Y.S. Swift, P.C. Hohenberg: Phys. Rev. **A15**, 319 (1977)
4.27 H. Haken: *Advanced Synergetics, Instability Hierarchies of Self-Organizing Systems and Devices*, (Springer, Berlin, Heidelberg, New York 1983)
4.28 K. Stewartson, J.T. Stuart: J. Fluid Mech. **48**, 529 (1971)
4.29 Y. Kuramoto: *Chemical Oscillations. Waves and Turbulence*, (Springer, Berlin, Heidelberg, New York 1984) p. 365

4.30 A.V. Gaponov-Grekhov, M.I. Rabinovich: Izv. VUZov, Radiofizika **30**, 131 (1987)

4.31 A.B. Ezersky, M.I. Rabinovich, V.P. Reutov, I.M. Starobinets: Sov. Phys. JETP **64**, 1228 (1986)

4.32 A.R. Bishop, J.C. Eilbeck, I. Satija, G. Wyzin: in Lect. Notes Appl. Math., Vol. 23 (Am. Math. Soc., New York 1986) p.72

4.33 G. Tesauro, M.C. Gross: Phys. Rev. **A34**, 1363 (1986)

4.34 H. Haken: Phys. Scripta **T9**, 111 (1985)

4.35 P. Coulett, C. Elphick, L. Gil, Y. Lega: Phys. Rev. Lett. **59**, 884 (1990)

4.36 A.V. Gaponov-Grekhov, A.S. Lomov, G.V. Osipov, M.I. Rabinovich: in *Nonlinear Waves 1*, A.V. Gaponov-Grekhov, M.I. Rabinovich, J. Engelbrecht (Eds.) (Springer, Berlin, Heidelberg, New York 1989)

4.37 M. Roberts, J.W. Swift, D.H. Wagner: Contemp. Math. **56**, 283 (1986)

4.38 M.S. Heutmaker, J.P. Gollub: Phys. Rev. **A35**, 242 (1987)

4.39 N.D. Mermin: Rev. Mod. Phys. **51**, 591 (1979)

4.40 V. Croquette, A. Pocheau: in *Cellular Structures in Instabilities*, Lect. Notes Phys., Vol. 210, J.E. Wesfreid, S. Zaleski (Eds.) (Springer, Berlin, Heidelberg, New York 1984) p. 104

4.41 O.V. Vashkevich, A.V. Gaponov-Grekhov, A.B. Ezersky, M.I. Rabinovich: Dokl. AN SSSR **90**, 960 (1986)

4.42 A.V. Gaponov-Grekhov, A.S. Lomov, M.I. Rabinovich: Pis'ma Zh. Eksp. Teor. Fiz. **44**, 224 (1986)

4.43 I.S. Aranson, A.V. Gaponov-Grekhov, M.I. Rabinovich: Izv. AN SSSR, ser. fiz. **51**, 1133 (1987)

4.44 P. Christiansen, P. Lombdal, N. Zabusky: Appl. Phys. Lett. **39**, 170 (1981)

4.45 I.V. Barashenkov, A.D. Gochera, V.G. Makhankov, I.V. Puzinin: Physika **D34**, 240 (1989)

4.46 A.S. Lomov, M.I. Rabinovich: Pis'ma Zh. Eksp. Teor. Fiz. **48**, 589 (1988)

4.47 K.A. Gorshkov, A.S. Lomov, M.I. Rabinovich: Phys. Lett. **A137**, 250 (1989); J.S. Aranson, K.A. Gorshkov, A.S. Lomov, M.I. Rabinovich: Physica **D43**, 435 (1990)

4.48 W. Jahnke, C. Henze, A.T. Winfree: Nature **336**, 662 (1988)

4.49 I.S. Aranson, M.I. Rabinovich: Izv. VUZov, Radiofizika **29**, 1514 (1986)

4.50 H. Linde: in *Self-Organization. Autowaves and Structures Far From Equilibrium*, V.I. Krinsky (Ed.) (Springer, Berlin, Heidelberg, New York 1984) p. 154

4.51 M.L. Smoes: in Proc. Int. Symp. on Synergetics, H. Haken (Eds.) (Springer, Berlin, Heidelberg, New York 1980) p. 80

4.52 S.P. Novikov, A.T. Fomenko: *Elements of Differential Geometry and Topology*, (Nauka, Moscow 1987)

4.53 I.S. Aranson, M.I. Rabinovich: Physica Math. and Gen. **A.23**, 286 (1990)

4.54 J.J. Hopfield: Proc. Math. Acad. Sci. USA **79**, 2554 (1982)

4.55 A.N. Kolmogorov: Usp. Mat. Nauk **38**, 27 (1970)

4.56 R. Abraham: *Simulation by Video Feedback*, Lect. Notes Math., Vol. 525 (Springer, Berlin, Heidelberg, New York 1976) p. 10

4.57 G. Ferrano, G. Häusler: Opt. Eng. **19**, 442 (1980)

4.58 J.P Crutchfield: Physica **D10**, 229 (1984)

4.59 V.N. Golubev, M.I. Rabinovich, V.I. Talanov: Pis'ma Zh. Eksp. Teor. Fiz. **42**, 84 (1985)

4.60 S.A. Zhukov, V.V. Barelko, A.G. Merzhanov: Dokl. AN SSSR **245**, 94 (1979)

4.61 V.A. Vasil'ev, Yu.M. Romanovsky, V.G. Yakhno: Usp. Fiz. Nauk, **128**, 625 (1979)

4.62 V.V. Barelko, Yu.E. Volodin: Dokl. AN SSSR, **223**, 112 (1975)

4.63 Ya.B. Zel'dovich, G.I. Barenblatt, V.B. Librovich, G.M. Mihviladze: *Mathematical Theory of Combustion and Explosion*, (Nauka, Moscow 1980)

4.64 S.A. Akhmanov, M.A. Vorontsov, V.U. Ivanov: Pis'ma Zh. Eksp. Teor. Fiz. **47**, 611 (1988)

4.65 G.E. Hinton, J.A. Anderson, N.Y. Earlbaum (Eds.) *Parallel Models of Associative Memory* T. Lawrence (Ed.), (Erlbaum Ass. Hillsdale 1981) pp.105-143

Chapter 5

5.1 V. Frost, F. Moulden: *Handbook of Turbulence. Fundamentals and Application*, Vol. 1 (Plenum, New York 1977)

5.2 L.D. Landau: Dokl. AN SSSR **44**, 339 (1944)

5.3 E. Hopf: Comm. Pure Appl. Math. **1**, 303 (1948)

5.4 L. Brandstätter, H.L. Swinney, A. Wolf, J.D. Farmer, E. Yen, P. Crutchfield: Phys. Rev. Lett. **51**, 1442 (1983)

5.5 B. Malraison, P. Atten, P. Berge, M. Dubois: *An Experimental Determination for the Chaotic Regime of Two Convective Systems*, Preprint, C.E.A. Saclay [Serv. Phys. Solide et de Rés. Magn.], Paris, France (1983)

5.6 A.S. Newell, J.A. Whithead: J. Fluid Mech. **36**, 239 (1969)

5.7 V.L. Ginzburg, L.D. Landau: Zh. Eksp. Teor. Fiz. **20**, 1064 (1950)

5.8 L.A. Segel: J. Fluid Mech. **38**, 203 (1969)

5.9 R. Graham: in *Fluctuations, Instabilities and Phase Transitions*, T. Riste (Ed.) (Plenum, New York 1975) p. 215

5.10 I.S. Aranson, A.V. Gaponov-Grekhov, M.I. Rabinovich: Physica **D33**, 1 (1988)

5.11 I.S. Aranson, A.V. Gaponov-Grekhov, M.I. Rabinovich: Zh. Eksp. Teor. Fiz. **89**, 92 (1985)

5.12 I.S. Aranson, A.V. Gaponov-Grekhov, M.I. Rabinovich, A.V. Rogal'sky, R.Z. Sagdeev: *Lattice Models in Nonlinear Dynamics of Nonequilibrium Media*, Preprint, Inst. Appl. Phys. USSR Ac. Sc., Nizhnii Novgorod, USSR (March 1987)

5.13 A.B. Ezersky, P.I. Korotin, M.I. Rabinovich: Pis'ma Zh. Eksp. Teor. Fiz. **41**, 129 (1985)

5.14 M.D. Gabovich, V.Va. Poritskii: Zh. Eksp. Teor. Fiz. **85**, 146 (1983)

5.15 A.B. Ezersky, M.I. Rabinovich, V.P. Reutov, I.M. Starobinets: Zh. Eksp. Teor. Fiz. **90**, 2051 (1986); M.I. Rabinovich, V.P. Reutov, A.V. Rogal'skii: Phys. Lett. **144A**, 259 (1990)

5.16 A.V. Gaponov-Grekhov, M.I. Rabinovich: Sov. Sci. Rev. A, Phys. **10**, 257 (1988)

5.17 A.V. Gaponov-Grekhov, M.I. Rabinovich, I.M. Starobinets: Dokl. AN SSSR, **279**, 596 (1984)

5.18 A.V. Gaponov-Grekhov, A.S. Lomov, G.V. Osipov, M.I. Rabinovich: in *Nonlinear Waves 1*, A.V. Gaponov-Grekhov, M.I. Rabinovich, J. Engelbrecht (Eds.) (Springer, Berlin, Heidelberg, New York 1989)

5.19 V.V. Kozlov, M.I. Rabinovich, M.P. Ramazanov, A.M. Reiman, M.M. Sushchik: Phys. Lett. **A128**, 479 (1988)

5.20 I.S. Aranson, M.I. Rabinovich: in Proc. Intl. Conf. Renormalization Groups in 1986, (World Scientific, Singapore 1987) p. 189

5.21 A.V. Gaponov-Grekhov, M.I. Rabinovich, I.M. Starobinets: Zh. Eksp. Teor. Fiz. **39**, 561 (1984)

5.22 V.S. Afraimovich, A.N. Verichev, M.I. Rabinovich: Izv. VUZov, Radiofizika **29**, 1050 (1987)

5.23 D.R. Moore, J. Toomre, E. Knobloch, N.O. Weiss: Nature **303**, 663 (1983)

5.24 K.N. Chic, G.A. Oswald, V. Chic: in Lect. Notes Eng., Vol. 24 (Springer, Berlin, Heidelberg, New York 1986) p. 118

5.25 R. Grappin, J. Leorat: Phys. Rev. Lett. **50**, 1100 (1987)

5.26 R. Grappin, J. Leorat, A. Pouquet: J. Physique **47**, 1127 (1986)

Acknowledgements of the Figures

Legal notice: For illustrations not stemming from the authors (and their collaborators) themselves the copyright for the use of these illustrations in a different context is with the colleagues named and/or their publishers.

The authors would like to thank them very much for their help and cooperation. The reader will certainly appreciate that the combined efforts of all these colleagues did help to arrive at a rather rare collection of beautiful illustrations.

Plate 1. Illustration courtesy of H.-O. Peitgen, P.H. Richter; *The Beauty of Fractals* (Springer, Berlin, Heidelberg, New York 1986) p.51
Plate 2. Illustration courtesy of H.-O. Peitgen, P.H. Richter; ibid, maps 11–13 p. 48
Plate 3. Illustration courtesy of H.-O. Peitgen, P.H. Richter; ibid, maps 43–44 p. 85
Plates 4–6. Illustrations courtesy of H.-O. Peitgen, P.H. Richter; ibid, maps 79, 83, 86 on pp.118–119
Plate 7. Illustration courtesy of C. Pezeshki, E.H. Dowell; Physica *D32*, 194 (1988) Plates XI to XIV
Plate 8. Illustration courtesy of H.-O. Peitgen, P.H. Richter; ibid, maps 31–33 on p.80
Plate 9. G.V. Osipov, M.I. Rabinovich; Inst. Appl. Phys., Russian Acad. Sci., Nizhny Novgorod, Russia
Plates 10, 11. Illustration courtesy of J.M. Ottino Scientific American *260*, 34 (1989)
Plate 12. Illustration courtesy of L.-D. Chen (Mead Imaging) and of Po-zen Wong; Physics Today *41*, 24 (1988)
Plate 13. Illustration courtesy of P. Meakin (E.I. du Pont de Nemours and Company, Inc.) and of L.M. Sander; Scientific American **256**, 82 (1987)
Plate 14. A.V. Rogalsky and M.I. Rabinovich; Inst. Appl. Phys., Russian Acad. Sci., Nizhny Novgorod, Russia
Plates 15–17. A.B. Ezerskii and M.I. Rabinovich; Inst. Appl. Phys., Russian Acad. Sci., Nizhnii Novgorod, Russia
Plates 18–20. Illustration courtesy of W.M. Roquemore, R.L. Britton, R.S. Tankin, C.A. Boedicker, M.M. Whitaker, and D.D. Trump; Phys. Fluids **31**, 2385 (1988)
Plate 21. P.R. Gromov, M.I. Rabinovich, M.M. Sushchik and A.B. Zobnin; Inst. Appl. Phys., Russian Acad. Sci., Nizhny Novgorod, Russia
Plates 22–23. A.B. Ezerskii and M.I. Rabinovich; Inst. Appl. Phys., Russian Acad. Sci., Nizhny Novgorod, Russia
Plate 24. A.B. Ezerskii and M.I. Rabinovich; Inst. Appl. Phys., Russian Acad. Sci., Nizhny Novgorod, Russia
Plates 25–28. A.B. Ezerskii, M.I. Rabinovich, V.P. Reutov; Inst. Appl. Phys., Russian Acad. Sci., Nizhny Novgorod, Russia
Plates 29. A.V. Rogalsky and M.I. Rabinovich; Inst. Appl. Phys., Russian Acad. Sci., Nizhny Novgorod, Russia
Plates 30–32. Illustrations courtesy of L.M. Volkov; LenNauchFilm
Plates 33–35. Illustration courtesy of L.M. Volkov LenNauchFilm
Plates 36–37. Illustrations courtesy L.M. Volkov LenNauchFilm
Plate 38. Illustration courtesy of H. Fiedler and M. van Dyke, ibid.
Plate 39. Illustration courtesy of R. Drabka, H. Nagib and M. van Dyke, ibid.

Plates 40–42. Illustrations courtesy of R.R. Prasad and K.R. Sreenivasan; Phys. Fluids *31*, 2393 (1988)
Plates 43–56. V.G. Yakhno and M.I. Rabinovich; Inst. Appl. Phys., Russian Acad. Sci., Nizhny Novgorod, Russia
Plates 57–62. V. Tolkov, V.G. Yakhno and M.I. Rabinovich; Inst. Appl. Phys., Russian Acad. Sci., Nizhny Novgorod, Russia
Plate 63. See Plate 9
Fig. 1. Illustration courtesy of E.N. Pelinovsky; Inst. Appl. Phys., Russian Acad. Sci., Nizhny Novgorod, Russia
Fig. 3. G.V. Osipov, M.I. Rabinovich; Inst. Appl. Phys., Russian Acad. Sci., Nizhny Novgorod, Russia
Fig. 4. Illustration courtesy of T. Vicsek of the Research Institute for Technical Physics of the Hungarian Academy of Sciences in Budapest and of L.M. Sander; Scientific American *256*, 82 (1987)
Fig. 5. Illustration courtesy of Ming-Yang-Su and M. van Dyke, Ed. *An Album of Fluid Motion* (The Parabolic Press, Stanford 1982) Fig. 194
Fig. 6a–c. Illustration courtesy of N. Rulkov; Inst. Appl. Phys., Russian Acad. Sci., Nizhny Novgorod, Russia

Subject Index

Adiabatic invariant 16
Alekseev 59
Andronov 2, 16, 36, 39, 40, 46, 87, 105
Anosov 61
Amplitude equation 90
Arnold 40
Arnold's diffusion 63
Attractor 34
– Lorenz 65, 68, 80
– multi-dimensional 125
– – strange 127
– nontrivial 34
– simple 73
– strange 6, 64, 68, 72, 140
Autocatalytic chemical reaction 85, 105
Autowaves 29
Averaged equation 36
Averaging method 14

Basin of attraction 35
Belousov-Zhabotinskii reaction 146
Bénard cell 86, 87
Bénard-Marangoni convection 92
Bifurcation 35, 40–46
– Andronov-Hopf 44
– local 44
– of equilibrium states 41, 42
– of the change of stability 41
– of periodic motion 41–43, 45
– period doubling 46, 149
– spatial 129
Bifurcation diagram 66
Bogolyubov 4
Boundary condition 28
– periodic 58
Boundary layer 120, 129
Bound state 104
Boussinesq equation 23, 25
Brillouin 48
Buck 149

Cantor structure 66, 76
Capacity 76, 78
Capillary gravity waves 27, 30
Cartwright 68
Cat's eye 19
Cellular processor 148
Chain
– of coupled nonlinear oscillators 21
– of generators 38
Chain model 122
Chaos 18, 136
– deterministic 117
– – spatio-temporal 123, 127, 146
– dissipative 64
– dynamical 134, 137
– Hamiltonian 64
Chaotic behaviour 60
Chaotic orbiting 60
Characteristic equation 44
Cnoidal wave 37
Combustion wave
Commensurability 127
Competition 36
– in space 50
– in time 50
– of modes 36, 44, 87
– of species 44
– of structures 93
– of waves 50, 124
Concentration structures 103
Conservative media 38
Convection 145
Convective instability 144
Correlation integral 82–84, 128
Couette-Taylor flow 110
Criterion
– Chirikov 62
– Melnikov 63
Critical phenomena 112
Critical point 73

Debye 48
Defect 92
– nonstationary 93
Degrees of freedom 12, 22, 32, 36, 119
– collective 31
– number of 4, 5, 36
Deterministic structure 6
Diffusion limited aggregation 143

Dimension 75, 77, 78
– correlation 128
– flow 127, 128
– fractal 76, 128
Disorder 85, 133, 146
Dispersion law 25
Dissipative media 29, 38
Dissipative system 32
– nonlinear 32
Dynamical chaos 134
Dynamic(al) system 7, 44
Dynamical stochastization 8
Dynamics of defects 95

Eigenfrequency 21
Eigenfunction 45
Einstein 48
Embryological approach 39
Energy conservation
– for coupled oscillators 15
Energy exchange 14
– quasi-periodic 22
Ensemble 7, 36
Entropy 75
– Kolmogorov 83
– topological 76
Equilibrium state 6, 8, 24, 28, 43, 44, 68, 69, 91
– saddle 28, 37
– stable 37, 104
Ergodicity 77
Evolution 28
– stable 138
– unstable 138
Excitable medium 112

Faraday 123
Faraday ripples 144–146
Fermi 14, 21, 23, 24
Fermi-Pasta-Ulama chain 24, 28
Fermi-Pasta-Ulama paradox 22, 23
Fermi-Pasta-Ulama problem 25
Feynman 117
Finite-dimensional model 4
Finite-dimensional system 39, 75
Fractal manifold 141
Fractals 139
Franken 16
Free energy functional 92

Gauze 13
Generator
– autonomous noise 69
– high-stability low-power 37
– noise 69, 71
– nonautonomous 37, 69
– self-excited noise 35
– stochastic 69
– van der Pol 33, 59, 69
Ginzburg 121
Ginzburg-Landau equation 124
– discrete version 38
– generalized 91, 121
– parametric analogue 91, 92
– two-dimensional 105, 131, 132
Gorelik 14
Gradient system 103
– generalized 103
Graham 60
Grassberger 82

Haken 60
Haken equation 90, 91
Hard excitation 35, 147
Helmholtz instablility 29
Hénon map
– with noise 83
Hénon-Heills system 17
Hexagonal lattice 144
Hierarchy of instabilities 118
Homoclinic structure 24, 25, 59, 62, 64, 68
Hysteresis 36

Image point 35
Incommensurability 127
Initial state 12
Instability
– collective 32
– explosive 12, 88
– modulation 53–55
– of trajectories 76
– parametric
Integrability 25
Intermittency 72–74, 130

Josephson junction 31

Kadomtsev-Petviashvili equation 30
Karman 118
Khokhlov 16
Kinetic equation 44
Klein-Gordon equation 26
Kolmogorov 108
Korteweg de Vries equation 23–26, 30
– modified 23
Kruskal 25, 26
Krylov 4

λ–ω-model 105
Landau 118, 121

Landau amplification 18
Landau damping 18
Landau-Hopf hypothesis 119
Landsberg 48
Langmuir waves 57
Lattice
– capillary 123
– of coupled pendulums 127
– periodic 123, 146
– two-dimensional
– – of generators 146
Lattice models 18
Ledrappier 80
Leontovich 16, 48, 53
Lighthill condition 53
Limit cycle 8, 35, 37, 43, 44
– unstable 35
Littlewood 68
Localization mechanism 100
Localized solution 30
Localized structures 101
– stationary 104
– two-dimensional 101
Lorenz 59
Lorenz system 68
Lotka-Volterra 139
Lyapunov 40
Lyapunov exponent 76, 78, 126, 131
– characteristic 79
Lyapunov functional 92, 93
Lyapunov theorem 29

Mandelbrot 139
Mandelstam 2, 12, 14, 24, 48
Mandelstam-Brillouin scattering 48, 49
Manley-Rowe relation 15
Memory 107
Metastable structure 103
Method of slowly varying amplitudes 34
Mode coupling 36
Mode locking 36
Modulation 46, 47, 51
– transverse 51
Modulation instability 53
Modulation waves 55
Multisoliton 25
Multistability 44, 92, 148
Mutual coupling 36

Navier-Stokes equation 117, 121, 131
Neuron-like media 113
Nonequilibrium medium 90, 100, 145
– dissipative 50, 101
Nonequilibrium thermodynamics 86
Nonlinear detuning 53
Nonlinear diffusion 90
Nonlinear dynamics 2, 149
Nonlinear Landau damping 20
Nonlinear optics 14
Nonlinear perturbation theory 29
Nonlinear polarization spectroscopy 51
Nonlinear Schrödinger equation 26, 53, 55–58
Normal mode 25

Observable 82, 84
Order 85, 133
Oscillation
– acoustic 21
– modulated 45, 47
– natural 37
– normal 21, 25
– quasi-harmonic 37
– quasi-periodic 37, 61
– self-excited 2, 8, 11, 31-40
– – collective 38
– – stochastic 64, 73, 74
– single-frequency 37
– transient 12
– undamped 32
Oscillator 18
– conservative 19
– coupled 13, 16, 18, 38
– – nonlinear 14, 16, 17
– harmonic 24
– low-frequency 15
– nonisochronous 20
– nonlinear 4, 8, 13, 17, 18, 20, 35, 38
– quantum mechanical 16
– quasi-harmonic 20, 36, 38
– self-excited 32, 70
– weakly nonlinear 16

Papaleksi 34
Particle-like property 26
Particle-like solution 101, 103
– three-dimensional 102
Particle-like structures 104
Parametrically coupled waves 49
Parametric instability 52, 53
Pasta 14, 21, 23, 24
Peach-Köhler's force 97
Peierls 3
Pendulum
– frictionless 19
– nonlinear 24
– spring 16
Period doubling 59
– infinite 72
Periodic trajectory 9
Phase conjugation 51
Phase plane 19, 25, 34
Phase portrait 60

Phase space 12, 34, 62
- effective 82, 83
- low-dimensional 62
- nonautonomous 35
- three-dimensional 17, 61
Phase synchronization 88
Phase trajectory 19
Phase volume 12, 64
Poincaré 8, 24, 25, 40
Poincaré limit cycle 8
Poincaré map 35, 61
Poincaré mapping 9, 45, 59, 80
Poincaré section 62
Poincaré time 119
Prandtl number 13
Procaccia 82

Quasi-linear physics 3, 14, 20, 47
Quasi-particles 3, 15, 56
Quasi-periodic motion 12, 15, 16
Quasi-periodic oscillations 46, 68

Raman 14, 49
Raman scattering 48, 49
Randomness 59, 63, 80, 134
Rayleigh 2
Rayleigh-Bénard convection 98, 121
Recurrence 4, 22, 23, 47, 54
Reverberator 86, 89
Reynolds 118
Reynolds number 118, 146
Ruelle 64
Running Mandelstam lattice 48

Scattering of solitons 26
Self-focussing 31
Self-focussing media 55
Self-modulation 31, 47, 48, 52
Self-organization 6, 86, 89, 134, 186
Self-organizing criticality 249
Self-structure 29, 87, 145
- convective 98, 145
Separatrix 7, 9, 13, 17, 19, 20, 25, 37, 62, 63
Shock waves 11, 27, 28, 35
- stationary 28, 29
Sinai 60
Sine-Gordon equation 92
- nonautonomous 92
Smale 60
Small perturbation 44
Soft onset 80
Solitons 6, 22, 23, 26, 27, 30, 31
- as particles 25
- dark 107
- dissipative 88
- field 56
- horseshoe-shaped 89
- modulation 56
- one-dimensional 30
- particle-like 22
- radio 57
- Rossby 29
- sine-Gordon 122
- three-dimensional 30, 57
- two-dimensional 30
- undisturbed 57
- video 56
Soliton gas 31
Spatial pattern 90
Spiral wave 105
Spontaneous symmetry breaking 44
Static pattern 107
Stationary wave 7
Stochastic behaviour 7
Stochastic dynamics 7, 60
Stochasticity 8, 24, 62, 72
Stochastic set 65, 75
Stokes wave 51
Structures 9
Supercriticality 127
Superposition principle 1
Swift-Hohenberg model 90
Synchronization 36
- external 36
- frequency 36
- mutual 36
Synergetics 86

Takens 64, 76, 81
Taylor 118
Taylor vortices 85, 122
Thermoconvection 60, 85
Three-dimensional particles 101
Three-dimensional spirals 105
Three-dimensional vortices 145
Time series 82
Tollmien-Schlichting waves 57
Topological equivalence 41
Torus
- m-dimensional 12
- two-dimensional 46, 76
Trajectory 12
Turbulence 11, 58, 117, 124, 146
- chemical 90, 146
- dynamical 146
- Langmuir wave 90
- weak 7, 23
TV analogue 110, 111

Ulam 14, 21, 23, 24

Van der Mark 69
Van der Pol 2, 3, 33, 36, 69
Van der Pol–Duffing system 69
Van der Pol equation 34, 39
Van der Pol nonautonomous system 68
Velocity distribution function 18
Vitt 5, 13, 14, 36
Vortex ring 105

Wave
– capillary gravity 30
Weakly nonlinear phenomena 3
Weakly supercritical convection 93
Williams 60

Zabusky 25, 26